S0-BZE-737

DOGFISH DISSECTION MANUAL

Bruce D. Wingerd

Illustrated by Geoffrey Stein, D.V.M.

THE JOHNS HOPKINS UNIVERSITY PRESS
Baltimore & London

© 1988 The Johns Hopkins University Press
All rights reserved
Printed in the United States of America

The Johns Hopkins University Press
701 West 40th Street
Baltimore, Maryland 21211
The Johns Hopkins Press Ltd., London

The paper used in this publication meets the minimum require-
ments of American National Standard for Information Sciences
—Permanence of Paper for Printed Library Materials, ANSI
Z39.48-1984.

ISBN 0-8018-3709-X (pbk.: alk. paper)

Contents

Illustrations

Introduction

THE SPINY DOGFISH is a small, cartilaginous fish that has long been a popular specimen for study in the field of comparative anatomy. Its popularity is due to its large numbers and consequent ease of capture in the sea, convenient storage, and represenative anatomy of primitive jawed fish. As a result, a wealth of information about its biology has been accumulated.

The purpose of this laboratory manual is to present a summary of the available information on the anatomy of the spiny dogfish. Its primary focus is the presentation of a logical and comprehendible sequence of dissection instructions that will guide the student through a pictorial journey of dogfish anatomy. The dissection procedures are supplemented by descriptions of basic functions, morphological adaptations, and structural relationships to other vertebrates.

The fossil record of the class of fish to which the spiny dogfish belongs dates back to the Devonian period, approximately 400 to 450 million years ago. The identifying characteristics that set the spiny dogfish apart from other fishes include jaws and paired fins, a cartilaginous skeleton, lateral gills that contain internal gill arches, dermal placoid scales, pectoral fins that are not unusual in size, a lack of anal fins, and the shape and arrangement of its teeth. Based on these structural features, the spiny dogfish is classified as follows:

Phylum: Chordata
Subphylum: Vertebrata
Superclass: Pisces
Class: Chondrichthyes
Subclass: Elasmobranchii
Order: Squaliformes
Family: Squalidae
Genus: *Squalus*
Species: *Squalus acanthias*

Before beginning your study of the dogfish, study the directional and spatial terms listed below. This introductory step is essential because these terms will be used extensively throughout the text (Fig. N.1).

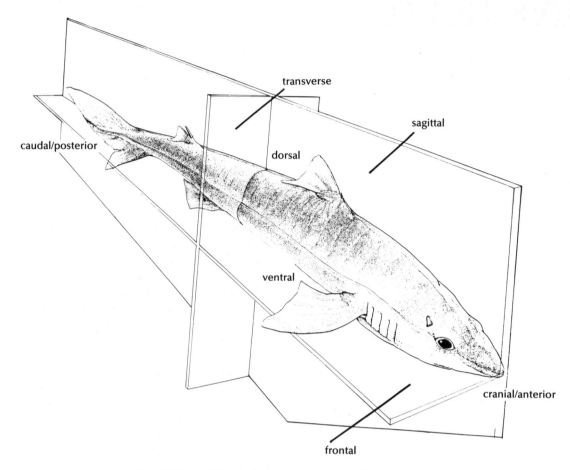

FIGURE N.1.   Descriptive terminology and planes of section

Directional Terms

Rostral: toward the nose end.

Cranial/anterior: toward the head end.

Caudal/posterior: toward the tail end.

Dorsal: toward the back side.

Ventral: toward the belly side.

Midline: an imaginary line that bisects the body into right and left halves.

Medial: lying closer to the midline relative to another structure.

Lateral: lying further from the midline relative to another structure.

Proximal: near a structure's origin or point of attachment to the body.

Distal: away from a structure's origin or point of attachment to the body.

Superficial: toward the body surface.

Deep: away from the body surface.

Planes of Section

Transverse (cross): a plane that passes at a right angle to the long axis of a body or body structure, usually resulting in cranial and caudal portions.

Longitudinal: a plane that extends from cranial to caudal along the long axis of the body; the longitudinal plane bisects the transverse plane at a right angle.

Sagittal: a longitudinal plane that divides the body into lateral right and left parts; if this division is into equal halves, it is called **midsagittal**. If it is into unequal parts, it is called **parasagittal**.

Frontal (coronal): a longitudinal plane that extends from cranial to caudal and horizontally from right to left, dividing the body into ventral and dorsal portions.

External Anatomy

1

THE EXTERNAL ANATOMY of the dogfish is representative of the shape and form of aquatic vertebrates. Its features are not parts of a single organ system but are manifestations of numerous systems that lie deep, with the exception of the skin, or **integument**. When combined as a unit, the external features of the dogfish help provide it with the ability to survive in its environment. In this chapter you will examine the external features of the shark, which include the integumentary system.

GENERAL EXTERNAL FEATURES

Place your specimen on a dissecting tray and examine its external features (Fig. 1.1). Note that its shape is streamlined, or **fusiform**, enabling it to glide easily through the water with the least possible resistance. The body is divided into three anatomical regions: the **head**, which extends from the tip of the snout to the pectoral fins; the **trunk**, which continues to the origin of the tail; and the **tail**, which is at the posterior end.

HEAD

Identify the following features of the head region (Fig. 1.1):

Nares: located ventral to the snout on either side. Inspect these paired openings closely, and you will observe each is divided into two channels: a rostral **incurrent aperture**, which receives incoming water, and a more caudal **excurrent aperture**, through which water exits. Within the nares is a small chamber, the **olfactory sac**, which contains olfactory nerve endings in its walls that are involved in the sense of smell.

Mouth: the crescent-shaped opening on the ventral side of the head. It is bordered laterally by deep **labial pockets**.

Eyes: the two large, laterally placed eyes are set in deep sockets in the head. Bordering the edges of the sockets are immovable **eyelids**. Continuous with the upper and lower eyelid is a tough membrane called the **conjunctiva**, which covers the eyeball.

1

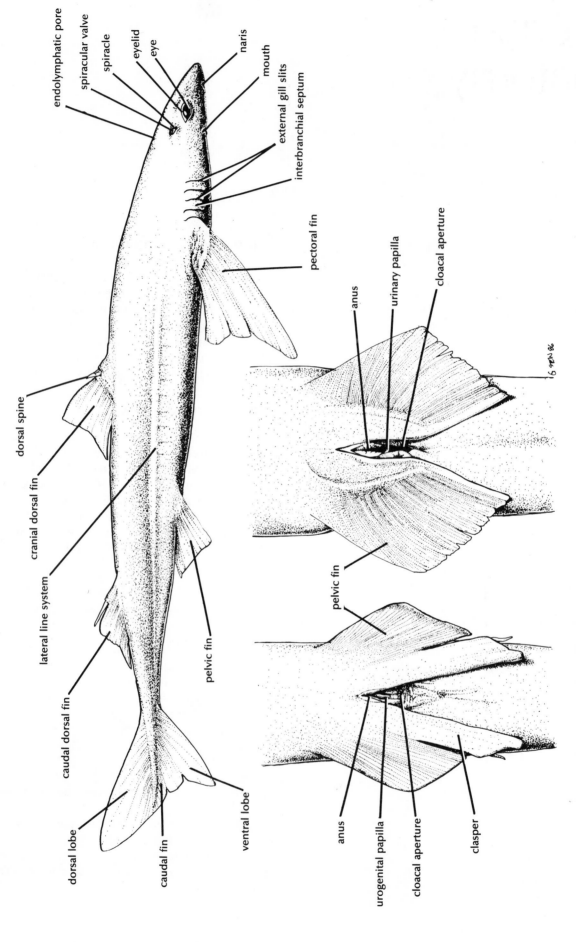

endolymphatic pore
spiracular valve
spiracle
eyelid
eye
naris
mouth
external gill slits
interbranchial septum

dorsal spine
cranial dorsal fin
lateral line system
caudal dorsal fin
dorsal lobe
caudal fin
pelvic fin
ventral lobe

pectoral fin

anus
urinary papilla
cloacal aperture
pelvic fin

anus
urogenital papilla
cloacal aperture
clasper

FIGURE 1.1.   Top: external features of the dogfish, lateral view. Lower left: pelvic region of the male, ventral view.
Lower right: pelvic region of the female, ventral view.

2

Spiracles: a pair of large openings caudal to the eyes. They are the reduced, nonfunctional first gill slits. Probe into this opening, and you will find it opens into the mouth cavity. The dogfish is able to bring water to its gills via the spiracles for respiration when the mouth is closed. For bottom-dwelling sharks, skates, and rays, this is the primary means of supplying the gills with fresh water. The fold of tissue partially covering the opening, or **external spiracular pore**, is the **spiracular valve**.

External gill slits: located between the mouth and the pectoral fins, they number five on each side. Each two adjacent slits are separated by an **interbranchial septum**. Water entering from the mouth and spiracles passes into the **pharynx**, the **internal gill slits**, and finally the **branchial chambers** that contain the **gill filaments**. Water exits through the external gill slits.

Endolymphatic pores: a pair of small openings on the top of the head between the spiracles on each side of the midline. They communicate with the inner ear, which senses sound and equilibrium.

Ampullae of Lorenzini: small openings into the portion of the **lateral line system** that extends to the head. The lateral line system detects vibrations of low frequency in the water, but the ampullae of Lorenzini are now thought to be modified for electroreception.

TRUNK

Examine the trunk of the dogfish as shown in Figure 1.1, and identify the following:

Dorsal fins: The dogfish has two fins on its dorsal side, cranial and caudal. Each dorsal fin has a large **spine** in front of it. These spines are for defensive purposes, and each is associated with a modified skin gland at its base that secretes a poison.

Pectoral fins: the paired fins near the dogfish's head region. They are in a more cranial position than those of other sharks. The pectoral fins function as rudders during swimming to enable the dogfish to change direction.

Pelvic fins: paired fins at the caudal end of the trunk on the ventral side. If your specimen is a male, its copulatory organs may be seen on the medial side of the pelvic fins. These fingerlike structures are called **claspers** and are used for the transfer of sperm to the female during mating. The pelvic fins are used as stabilizers during swimming.

Cloacal region: The **cloaca** is the chamber within the dogfish's body that receives solid waste products from the intestinal tract, liquid waste from the urinary tract, and reproductive cells from the genital tract. It opens at the surface by way of the **cloacal aperture**, which may be seen between the two pelvic fins on the ventral side. From the cloaca, liquid urine passes through ducts to exit through an opening at the tip of the **urinary papilla**

(in females) or the **urogenital papilla** (in males), which can be seen within the cloacal aperture. Also within the cloacal aperture and lying cranial to the urinary or urogenital papilla is the **anus**, which is the opening of the intestinal tract into the cloaca.

TAIL

The tail of the dogfish is a large, powerful structure that is the primary source of locomotion. At its caudal end is the **caudal fin**, which consists of two portions: a larger **dorsal (epichordal) lobe** that lies above the vertebral axis, and a smaller **ventral (hypochordal) lobe** that lies below. This asymmetrical shape is characteristic of primitive fishes and is called **heterocercal**. It results in a greater thrust being provided by the dorsal lobe. In fact, if the stabilizing influence of the lateral fins and the flattened head were not available, the shark would tend to somersault. This tail structure is in contrast with the symmetrical **homocercal** caudal fin of the more advanced bony (teleost) fishes.

THE INTEGUMENT

The integument of the dogfish consists of two distinct layers: a superficial **epidermis** and a more extensive, deep **dermis**. If prepared slides of shark skin are available, examine them under approximately 100 magnification, and identify the following features of the epidermis and dermis (Fig. 1.2).

EPIDERMIS

The epidermis is composed of living cells and is without a keratinized outer layer found in most terrestrial vertebrates. This condition permits an osmotic exchange between the dogfish and the water medium. The basement layer of cells is called the **basal**, or **germinative**, layer; these cells are columnar in shape. As the cells near the body surface, they become progressively flatter in shape except for the enlarged mucous cells.

DERMIS

The dermis is primarily composed of dense connective tissue. Near its border with the epidermis are scattered pigmented cells called **chromatophores**. Chromatophores are also present in the dermis of fishes, amphibians, and reptiles and contain pigment that provides the animal with protective coloration.

Also located in the superficial portion of the dermis is the base of the **placoid scale**. Placoid scales extend through the epidermis to the exterior. The exposed portion of the scale is the **crown** and contains a **central spine** and two **lateral spines**. The portion that passes through the epidermis is the **neck**, which contains a central **pulp**

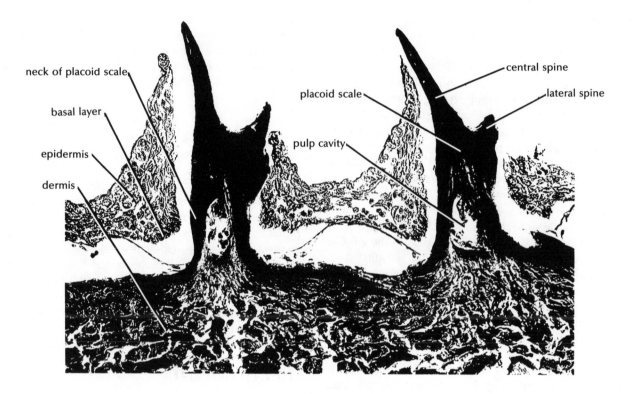

neck of placoid scale

basal layer

epidermis

dermis

central spine

placoid scale

lateral spine

pulp cavity

FIGURE 1.2.   Transverse section of skin, 400 ×. The space between the epidermis and dermis is an artifact made during sectioning.

cavity. A section through a placoid scale reveals an internal structure that is similar to that of a tooth: surrounding the pulp cavity is a thick layer of calcified **dentin**, and external to the dentin in the crown region is a hard layer of **enamel**. Placoid scales in the shark are nearly all that remain of the **dermal skeleton**, which once made up the extensive bony armor of more primitive fishes.

With a hand lens, examine the surface of the dogfish skin, and note the distribution of the placoid scales (chromatophores may also be seen as small, dark spots). Now slide your finger along the skin surface to detect the rough texture created by the presence of the scales. Is it any wonder that shark skin was once used as an abrasive for polishing wooden furniture!

The Skeletal System

THE SKELETON of the dogfish is primarily an **endoskeleton**, for it is an internal structure that arises from areas of the body deep to the skin. The other type of skeleton that makes up the supporting framework of vertebrates is the **dermal skeleton**. This has been reduced in the dogfish to consist of only the placoid scales and the teeth, both of which arise from the dermis. The endoskeleton consists of two basic portions: the **visceral skeleton**, which supports the gills and contributes to the structure of the jaws, and the **somatic skeleton**, which consists of all remaining endoskeletal components. The somatic skeleton is further divided into an **axial division**, which is composed of skeletal units lying on or near the median axis, and an **appendicular division**, which consists of the skeleton of the paired appendages and their supporting attachments.

The endoskeleton of the dogfish is composed entirely of cartilage. The cartilage of the dogfish skeleton is the same type of tissue that precedes bone formation in vertebrates that have a bony skeleton. Unlike bone, it is, for the most part, a delicate material that is easily damaged. It will therefore not be observed through dissection except for portions of the vertebral column. Instead, you are asked to study the skeleton of the shark using models and preserved mounts that may be available in the lab and the drawings in this text as a reference. Before proceeding, study the complete skeleton of the dogfish in Figure 2.1 to orient yourself to its general organization.

THE SKELETON OF THE HEAD REGION

The skeleton of the dogfish's head is composed of two portions: the **chondrocranium** and the **splanchnocranium**. The chondrocranium is a cartilaginous skull that forms the anterior end of the axial skeleton. It surrounds most of the brain and forms protective capsules around the olfactory sacs and inner ears. The splanchnocranium is the only component of the visceral skeleton and forms the visceral arches that support the gills and jaws.

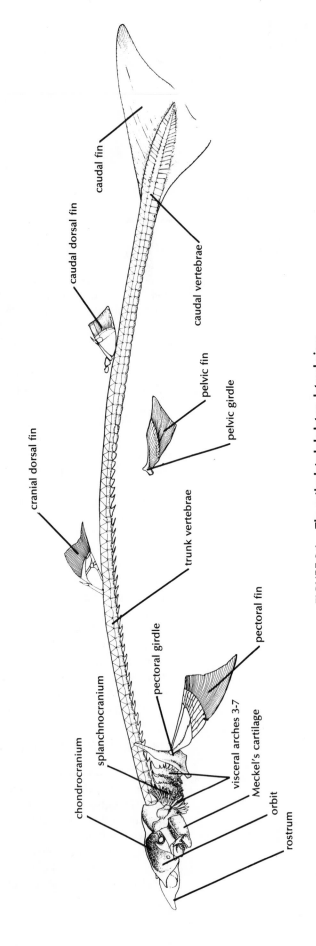

caudal fin

caudal dorsal fin

caudal vertebrae

cranial dorsal fin

pelvic fin

pelvic girdle

trunk vertebrae

pectoral girdle

pectoral fin

chondrocranium

splanchnocranium

visceral arches 3-7

Meckel's cartilage

orbit

rostrum

FIGURE 2.1.   The articulated skeleton, lateral view

6

CHONDROCRANIUM

The chondrocranium is a single unit of cartilage that is without divisions or sutures as seen in vertebrate skulls composed of bone. Examine the cartilaginous skull of the dogfish, and identify the following features (Figs. 2.2 and 2.3):

Rostrum: the anterior end that supports the snout. It contains the following:

    Precerebral cavity: a large depression on the dorsal side. It is filled with a gelatinous material in life.

    Rostral carina: On the ventral side, it is a keel-like midventral blade.

    Rostral fenestrae: a pair of openings into the cranial cavity near the base of the rostral carina. In life, these are closed by a membrane.

Nasal capsule: a pair of almost spherical structures located on each side at the base of the rostrum. They are often broken in preserved specimen due to their thin walls. On the ventral side, each capsule may be observed to contain two small holes, the **external nares**, which permit communication with the organs of smell within. The posterior wall of each capsule is continuous with the orbit by way of a ridge called the **antorbital process**.

Orbit: a pair of large, lateral depressions in the middle of the chondrocranium. In life each contains the eye, which is supported by a disk anchored to the medial wall of the orbit by a slender stalk. The supportive disk is called the **optic pedicle**. Identify the following features of the orbit:

    Supraorbital crest: A thick lateral shelf, it forms the dorsal wall of the orbit. Note the series of foramina on the dorsal side of the supraorbital crest. These are called **superficial ophthalmic foramina** and permit passage of branches from the trigeminal (V) and facial (VII) cranial nerves.

    Postorbital process: a lateral projection posterior to the supraorbital crest.

    Optic foramen: a hole through the medial wall of the orbit that permits passage of the optic nerve (II).

    Epiphyseal foramen: a single opening on the dorsal side just posterior to the base of the rostrum between the orbits. In life, the **epiphysis** or **pineal body** projects through. The epiphysis is a rudimentary third eye that was functional in many primitive vertebrates but now serves mainly endocrine functions.

    Basitrabecular process: a pair of lateral, rounded projections on the ventral side between the two orbits.

Otic capsule: a square region of cartilage posterior to the orbits. The paired otic capsules contain the inner ear of each side. Identify the following features present in this region:

Endolymphatic fossa: a large depression on the dorsal side between the two capsules. Within the fossa are two pairs of openings that communicate with the inner ear: the **endolymphatic foramina** and the **perilymphatic foramina**. The endolymphatic and perilymphatic ducts of the inner ear pass through these foramina.

Basal plate: the flattened surface on the ventral side between the otic capsules. Note the median white line that extends down the basal plate. This is the **notochord**.

Carotid canal: a small opening at the anterior end of the notochord. It permits passage of the internal carotid arteries into the cranial cavity.

Vagus nerve foramina: a pair of small openings at the posterior end of the otic capsules. They permit the vagus nerve (X) to exit from the cranial cavity.

Glossopharyngeal foramen: a pair of small openings lateral to the vagus nerve foramina at the caudodorsal angle of the chondrocranium. Each allows passage of the glossopharyngeal nerve (XI).

Occipital region: the posterior extremity of the chondrocranium. It contains the following:

    Foramen magnum: a large single hole that is visible on the dorsal side at the posterior end of the chondrocranium. The spinal cord passes through this hole to enter the cranial cavity.

    Occipital condyle: a pair of lateral projections at the posterior end of the basal plate that articulate with the first vertebra.

SPLANCHNOCRANIUM

The splanchnocranium or visceral skeleton is the part of the skull that forms the jaws, supports the gills, and provides attachment to respiratory muscles. In the dogfish it is composed of seven **visceral arches**. Obtain a specimen that contains an intact visceral skeleton and identify its following components (Fig. 2.4):

Mandibular arch: the first, or cranial-most, visceral arch. The mandibular arch forms the upper and lower jaws. It is composed of a dorsal and ventral half on each side:

    Palatoquadrate cartilage: Paired cartilages that form the dorsal half of the mandibular arch, they are attached at their medial ends. Together they constitute the upper jaw. A dorsal projection on each cartilage, called the **orbital process**, provides attachment to the ventral side of the chondrocranium by way of a ligament in the living fish. Note the multiple rows of **teeth** present.

    Meckel's cartilage: paired cartilages that form the ventral half of the mandibular arch. The Meckel's cartilages form the lower jaw and also bear multiple rows of teeth.

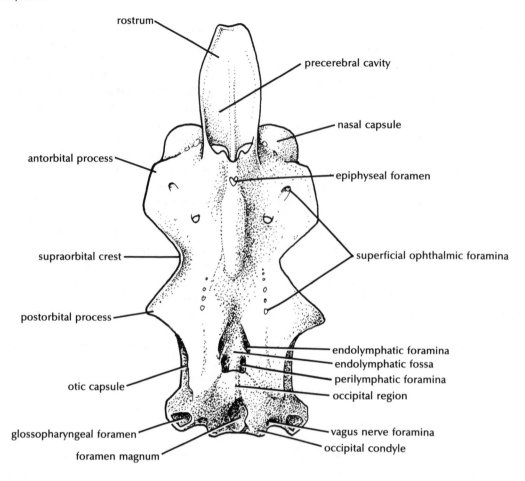

FIGURE 2.2. Chondrocranium, dorsal view

Hyoid arch: the second visceral arch. It extends ventrally from the otic capsule to provide support for the upper and lower jaws by way of ligaments. The hyoid arch is composed of two paired and one single cartilaginous elements:

>**Basihyal**: the single midventral portion of the hyoid arch.

>**Ceratohyal**: a pair of slender elements that extend from the basihyal to the angle of the jaw on each side. Note the slender **gill rays** on the posterior margins of the ceratohyals. These projections support the gills.

>**Hyomandibular:** a pair of short elements that extend dorsally and medially from their union with the ceratohyals. The hyomandibular articulates with the lateral surface of the otic capsule and the mandibular cartilage by means of ligaments, thereby providing support for the jaws.

Branchial arches: the remaining five pairs of visceral arches. The branchial arches support the gills. They bear numerous gill rays and short medial projections called **gill rakers**. They are relatively uniform in size, and each consists of the following four pairs of elements, from dorsal to ventral:

Pharyngobranchial: The dorsal-most elements, they project dorsocaudally.

Epibranchial: the elements ventral to the pharyngobranchials.

Ceratobranchial: the prominent middle elements.

Hypobranchial: ventral elements that project caudally. There are only three pairs of hypobranchials in the dogfish.

Basibranchial: ventral, or median, elements that are not paired. There are two basibranchials in the dogfish.

THE VERTEBRAL COLUMN

The vertebral column extends from its union with the chondrocranium into the tail. It protects the spinal cord and provides a point of attachment for trunk and caudal muscles. In the shark as in all fishes, it consists of two types of vertebrae: the **caudal vertebrae** and the **trunk vertebrae**.

CAUDAL VERTEBRAE

The caudal vertebrae may be studied by examining a prepared specimen or by dissection. If dissection is

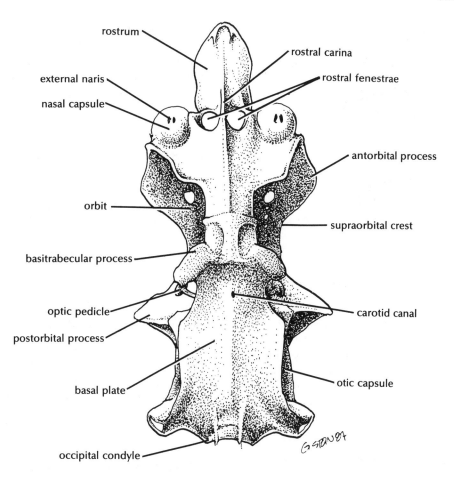

FIGURE 2.3.   Chondrocranium, ventral view

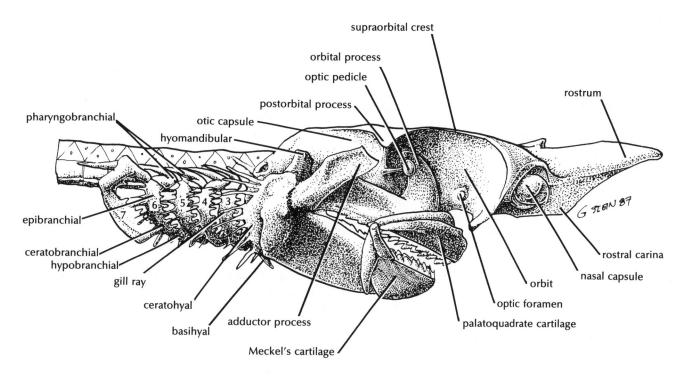

FIGURE 2.4.   Chondrocranium and splanchnocranium, lateral view

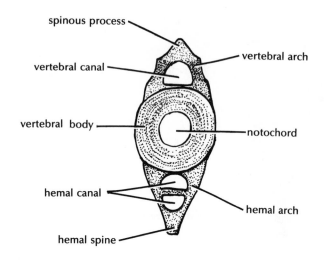

FIGURE 2.5. Transverse section through a caudal vertebra

chosen, begin by cutting completely through the vertebral column in the tail just caudal to the caudal dorsal fin. Make another cut 5 cm caudal from the first cut, and carefully expose the vertebrae by cleaning away the surrounding muscle and connective tissue. Now remove this section of vertebrae and make a clean cross section through the joint between two vertebrae. Identify its following features (Fig. 2.5):

Vertebral body: Also called the **centrum**, it is the cylindrical central portion.

Notochord: the primitive cord of vertebrates that has persisted from the embryonic stage in the shark. It lies within the vertebral body.

Vertebral arch: A dorsal arch projecting from the vertebral body, it protects the spinal cord. Its pointed tip is called the **spinous process**. Viewed laterally, the vertebral arch can be seen to consist of interdigitating triangular plates. These are called **dorsal intercalary plates** and **neural plates**. Each plate contains a small **nerve foramen** through which passes a spinal nerve root.

Vertebral canal: the cavity within the vertebral arch that contains the spinal cord.

Hemal arch: a ventral arch projecting from the vertebral body. Its pointed tip is called the **hemal spine**. The hemal arch is composed of a single plate of cartilage as viewed laterally, called the **ventral intercalary plate**.

Hemal canal: the cavity within the hemal arch; it contains the caudal artery and vein.

TRUNK VERTEBRAE

Examine the vertebrae of the trunk region by the use of a prepared specimen or dissection. If you are to dissect, follow the protocol above for removal of a caudal vertebra section in the trunk region and identify its components indicated in Figure 2.6.

Trunk vertebrae are similar to caudal vertebrae in many respects. However, note that trunk vertebrae do not have a hemal arch and hemal canal. In their place are a series of lateral processes, called **basapophyses**. The basapophyses are homologous with the proximal portion of the hemal arch and provide a point of attachment for the cartilaginous **ribs**.

Now examine the dorsal fins on a mounted skeleton. Each dorsal fin consists of fibrous rays called **ceratotrichia**, that are supported by several cartilages, collectively called **pterygiophores**. The cartilages are arranged in two series: the proximal series adjacent to the vertebral column are **basals**, and the more distal series are **radials**. Also note that the enlarged ceratotrichia of the caudal fin are not supported by pterygiophores as in the dorsal fins but by enlarged spinous processes and hemal spines from the vertebral column.

THE APPENDICULAR SKELETON

The appendicular skeleton consists of the skeletal material in the appendages and their attachments to the axial skeleton. In the dogfish this includes the **pectoral girdles**, **pectoral fins**, **pelvic girdles**, and **pelvic fins**. Identify the following components of the appendicular skeleton, using prepared specimens (Figs. 2.1, 2.7, and 2.8).

PECTORAL GIRDLE & APPENDAGES

The pectoral girdle consists of a U-shaped bar of cartilage immediately caudal to the gills. Identify the following components of the pectoral girdle and the attached pectoral fins (Fig. 2.7):

Coracoid bar: a large, U-shaped cartilage that encircles the ventral side of the trunk. It is the principle element of the pectoral girdle and articulates with each pectoral fin at the **glenoid surface** on each side of the trunk.

Scapular process: a segment of cartilage that is continuous with the coracoid bar and extends distally past the glenoid surface on each side of the trunk.

Suprascapular cartilage: a separate piece of cartilage that attaches to the distal end of each scapular process. Together with the coracoid bar and the scapular processes, they make up the pectoral girdle.

Basal fin cartilages: three cartilages at the base of each pectoral fin that articulate with the pectoral girdle at the glenoid surface. The basals are, from cranial to caudal, the **propterygium**, the **mesopterygium**, and the **metapterygium**.

Radial cartilages: numerous small cartilages that extend from the basal cartilages to the ceratotrichia. As in the dorsal fins, the basal and radial cartilages are collectively called **pterygiophores** and support the pectoral appendage.

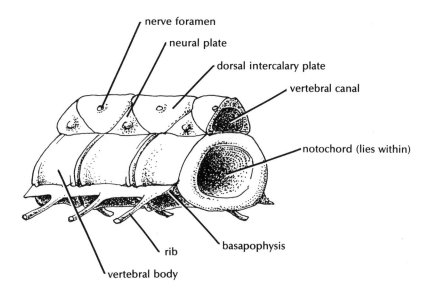

FIGURE 2.6.   Spinal column segment of the trunk region

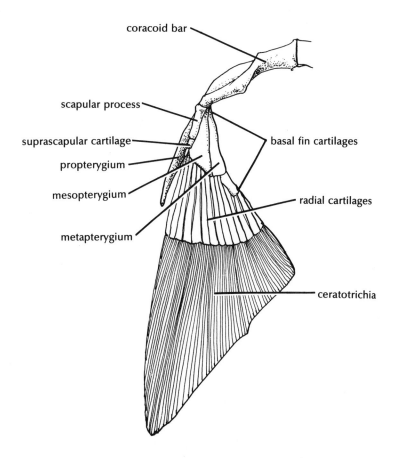

FIGURE 2.7.   Pectoral girdle and fin, lateral view

PELVIC GIRDLE & APPENDAGES

Examine the pelvic girdle and appendages in a preserved skeleton and identify the following components (Fig. 2.8):

Puboischiadic bar: a paired cartilaginous rod that makes up the pelvic girdle. It is located in the ventral abdominal wall cranial to the cloaca. At each end of the rod is a short process, the **iliac process**, which extends in a dorsal direction. Ventral to the iliac process is a smooth surface where the pelvic fins articulate with the girdle called the **acetabular surface**.

Basal fin cartilages: the pelvic appendages contain only two basals: a short **propterygium** that projects laterally from the girdle and a long **metapterygium** that extends caudally and forms the main support for the fin.

Radial cartilages: numerous small cartilaginous elements that extend from the basals to the ceratotrichia. In males, several radials are modified to form the skeleton of the long **clasper**, which is used for sperm transfer during mating.

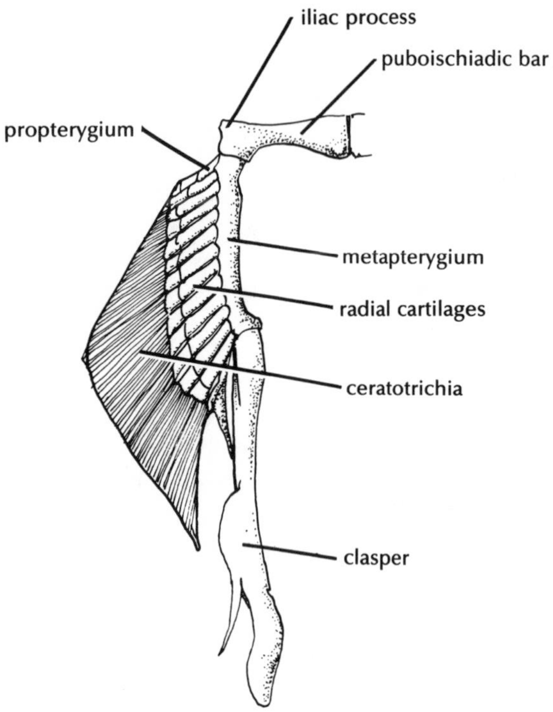

FIGURE 2.8. Pelvic girdle and fin of a male, lateral view

The Muscular System

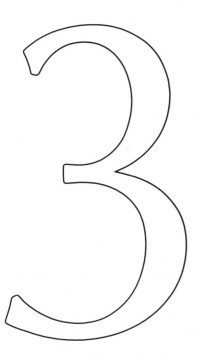

THE MUSCULAR SYSTEM consists of voluntary muscle, called **somatic** or **skeletal muscle**, and its associated connective tissue. Somatic muscle is embryonically derived from the "outer body tube" and is therefore the muscle type found in abundance immediately deep to the skin and attached to appendages. Because of this location, it can be thoroughly studied through dissection following removal of the skin.

In addition to somatic muscle, vertebrates also contain two other types of muscle tissue: **visceral muscle** and **cardiac muscle**. Visceral muscle forms the walls of visceral organs, blood vessels, and in fishes, the gills. During its embryonic development, it is associated with the "inner tube" of the body and, for the most part, is under involuntary control from the brain (the exception to this rule is the visceral muscles associated with the gills, called **branchial muscles**, which are voluntary). Cardiac muscle, as its name implies, is the muscle tissue that forms the bulk of the heart. It is also involuntary in its function. Because visceral and cardiac muscles are not part of the muscular system, they will not be studied at this time with the exception of the musculature of the gills. Instead, they will be studied when the body structure with which they are associated is examined.

The principal function of the muscular system is to provide skeletal movement. Movement is achieved by the shortening, or **contraction**, of muscles that are attached to skeletal elements. During contraction, one end of the muscle remains relatively stationary while the opposite end moves the skeletal element to which it is attached as the muscle shortens. The more stable end is called the **origin** of the muscle, and the mobile end is known as the **insertion**. Actual movement is produced by the force a muscular contraction exerts on its connective tissue attachments to the skeleton. The resulting body movement is termed **action**.

In this chapter you will study the muscular system of the dogfish, which is representative of most fishes. Each major muscle is listed according to its regional location and is identified by a description of its point of origin, point of insertion, and primary action. Dissection of the musculature of the dogfish is divided into somatic trunk and tail, fin, branchial, and hypobranchial regions.

SOMATIC MUSCLES OF THE TRUNK & TAIL

The somatic trunk and tail muscles make up the bulk of fish musculature. They form the walls of the trunk and occupy the tail region to the caudal fin. As such, they are the primary locomotors of the body.

To view them, remove a strip of skin from one side of the dogfish tail and trunk. This may be done by making a shallow incision about 5 cm in length with a scalpel through the skin along the dorsal midline, beginning behind the caudal dorsal fin and working caudad. Make a similar incision along the ventral midline. Connect these cuts by making transverse incisions on the left side from dorsal to ventral. Continuing with your scalpel, peel the skin away. Repeat this procedure in the trunk region to remove a second strip between the pectoral fins and the cranial dorsal fin.

Observe that the trunk and tail musculature are composed of vertical, zigzagged segments separated by white connective tissue septa (Figs. 3.1 and 3.2). The segments are called **myomeres**, and the connective tissue partitions are termed **myosepta**. The myomere fibers extend longitudinally, and the myosepta pass deep to attach to the vertebral column in a dorsoventral direction. Also note the white longitudinal line that extends along the middle side of the body. This is the **horizontal skeletogenous septum** and lies deep to the lateral line. It divides the myomeres into a dorsal, or **epaxial**, portion and a ventral, or **hypaxial**, portion.

The epaxial myomeres are arranged to form two or more **dorsal longitudinal bundles**, which extend along the trunk and tail regions. The hypaxial myomeres are composed of a darker **lateral longitudinal bundle** and two lighter **ventral longitudinal bundles**. The lateral bundle extends from the pectoral girdle to the tail, and the ventral bundles extend from the pectoral girdle to the pelvic girdle. The longitudinal bundles are visible in cross section (Fig. 3.3).

Also note the white partition along the ventral midline. It separates the hypaxial muscles of the two sides of the body into right and left and is called the **linea alba**.

FIN MUSCLES

The somatic muscles that move the fins are few in number and small in size, as they are used mainly for stability and not for locomotion by the dogfish. There are a pair of muscles associated with each of the four appendages. Each pair consists of a muscle mass located on the dorsal surface of a fin and one on the ventral surface.

MUSCLES OF THE PECTORAL FIN

Remove the skin from the base of the left pectoral fin by making an initial shallow incision and pulling the skin from the musculature. Continue this process on both surfaces of the fin and identify the following muscles (Figs. 3.1, 3.4, and 3.5):

Extensor: the dorsal muscle mass that extends into the fin from its base. It originates from the pectoral girdle and inserts on the pterygiophores and ceratotrichia. The extensor elevates the fin and pulls it caudally.

Flexor: the ventral muscle mass of the pectoral fin. It arises from the coracoid bar of the pectoral girdle and inserts on the pterygiophores. The flexor depresses the fin and pulls it cranially.

MUSCLES OF THE PELVIC FIN

Remove the skin from the left pelvic fin and its base as you had done with the pectoral fin. If your specimen is a female, this procedure is uncomplicated. If your specimen is a male, however, you will find a large muscular sac on the ventral surface of each pelvic fin called the **siphon**. The previously identified clasper will also be seen ventrally. Remove the siphon on the left side and leave it intact on the right. The siphon and clasper will be further studied later with the urogenital system. Identify the following muscles of the pelvic fin:

Extensor: As in the pectoral fin, this is the dorsal muscle mass. Its origin is from the surface of the caudal trunk myomeres, the iliac process, and the metapterygium. It inserts on the radial cartilages. The extensor elevates the pelvic fin.

Flexor: the ventral muscle mass; it is divided into proximal and distal segments. The proximal flexor arises from the puboischiadic bar and inserts on the metapterygium. The distal flexor arises from the metapterygium and inserts on the radial cartilages. Their combined action is depression of the pelvic fin.

BRANCHIAL MUSCLES

The branchial muscles are located in the head region and move the visceral arches and jaws. They are classified as visceral muscles because of their embryonic origin from the inner body tube, but they respond to voluntary control from the brain. Branchial muscles are divided into three muscle groups: the **superficial constrictors**, the **levators**, and the **interarcuals**.

SUPERFICIAL CONSTRICTORS

To examine the superficial constrictors, it will be necessary to skin the head region. Begin by making a shallow incision along the midventral line from the pectoral girdle to the end of the snout. Now make several cuts at right angles to the first cut and extend them to the mid-dorsal line. Cut around the gill region, the eyes, and the spiracles. Now strip the skin from the head by pulling it with your fingers from ventral to dorsal midlines until all skin is

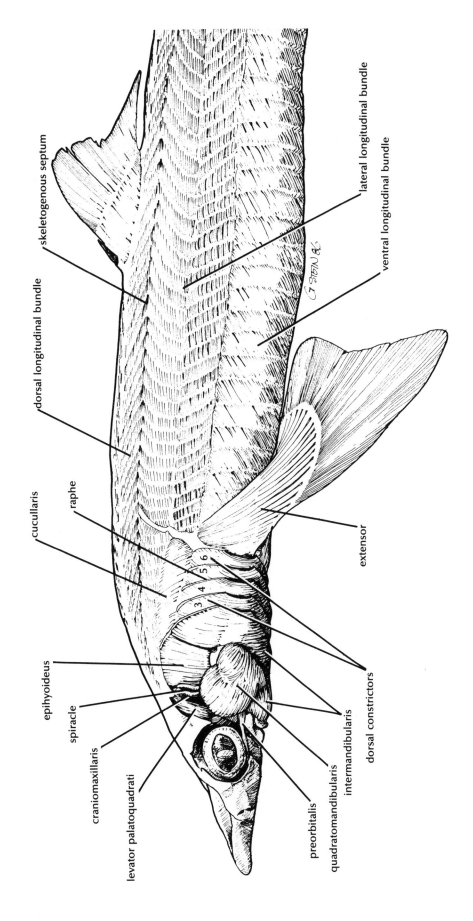

skeletogenous septum

dorsal longitudinal bundle

cucullaris

raphe

epihyoideus

spiracle

craniomaxillaris

levator palatoquadrati

preorbitalis

quadratomandibularis

intermandibularis

dorsal constrictors

3 4 5 6

extensor

ventral longitudinal bundle

lateral longitudinal bundle

G STEIN 96

FIGURE 3.1. Muscles of the head and trunk, lateral view

15

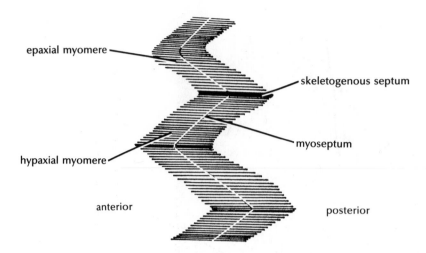

FIGURE 3.2. Schematic diagram of trunk myomeres, lateral view

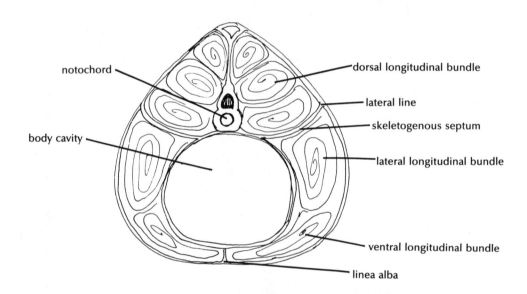

FIGURE 3.3. Schematic diagram of trunk muscles, transverse section

removed. Note that the dorsal longitudinal bundles of epaxial myomeres continue above the gill slits to attach to the caudal part of the chondrocranium. Also note the region cranial to the pectoral fins is not occupied by myomeres. The lateral muscles of this region are the branchial muscles. The ventral portion of this region is occupied largely by hypobranchial muscles (to be discussed later).

Superficial constrictor musculature covers the head and gill region from the pectoral fin to the eye and mouth. It consists of six **dorsal constrictors** located above the gill slits and six **ventral constrictors** below the gill slits. Each constrictor pair therefore consists of a dorsal portion and a ventral portion and corresponds to one of six visceral arches. The six pairs are separated from one another by a white vertical band of connective tissue called a **raphe**.

Their combined function is to compress the pharyngeal chamber, close the gill slits, close the mouth, and assist in opening the mouth. Identify the following superficial constrictor muscles (Figs. 3.1, 3.4, and 3.5):

First constrictor pair: the first constrictor corresponds to the first visceral arch, which forms the upper and lower jaws. As such, it is mainly involved in moving the jaws. Each constrictor pair is represented by the following four separate muscles:

    Craniomaxillaris (spiracularis): a small muscle that is immediately anterior to the spiracle. It originates on the optic capsule and inserts on the palatoquadrate cartilage. It elevates the palatoquadrate cartilage.

Quadratomandibularis (adductor mandibulae): a large round muscle ventral to the spiracle. Its origin is on the posterior part of the palatoquadrate cartilage and, its insertion is on Meckel's cartilage. The quadratomandibularis closes the mouth.

Preorbitalis: a small slender muscle between the upper jaw and the eye. This muscle will be traced later to avoid damage to the orbit. The origin of the preorbitalis is on the midventral surface of the chondrocranium, and it inserts onto fibers of the large quadratomandibularis.

Intermandibularis: a broad muscle that constitutes the ventral portion of the first constrictor. It lies posterior to the mouth and originates from a midventral raphe. From here it extends to the Meckel's cartilage and fascia of the quadratomandibularis where it inserts. The intermandibularis elevates the floor of the mouth.

Second constrictor pair: the widest and anterior most of the second through sixth constrictor pairs, it extends from the second gill slit to the angle of the jaw. It compresses the branchial pouches and contains two subunits:

Epihyoideus: an anterior continuation of the main portion of the second constrictor, it is located dorsal to the quadratomandibularis. Its origin is from the otic capsule and surrounding fascia, and it inserts on the hyomandibular.

Interhyoideus: a deep muscle beneath the intermandibularis near the ventral midline. To expose it, carefully cut the intermandibularis slightly to one side of the ventral midline. Make a very shallow cut to prevent damage to deeper muscles. Now make a second longitudinal cut from the angle of the jaw back two inches and connect the two cuts with right angle cuts. Remove the intermandibularis on the one side and observe the deep interhyoideus. It originates from the midventral raphe and inserts on the certohyal. Both the epihyoideus and interhyoideus compress the branchial pouches.

Third through sixth constrictor pairs: the muscles associated with the third through sixth visceral arches. Note that they overlap each other so that the more anterior constrictor partially conceals the one adjacent to it. The fibers of these superficial constrictors attach to the connective tissue raphe and extend toward the gill slits. Their contraction compresses the pharyngeal chamber and closes the gill slits. Concealed deep to the constrictors are muscles that form part of the gill structure. To expose them, begin to open the branchial pouches by cutting through the superficial constrictors in a direction perpendicular to the vertebral axis. Open them as wide as possible so that you can view the **interbranchial septa**, which support the **gill filaments** and their secondary foldings, the **primary lamellae** (refer to Figs. 5.2 and 5.3). Remove the skin and gill filaments from the anterior face of one septum and note the circular arrangement of muscle fibers forming the wall. This deep muscle is the **interbranchial**. There are four interbranchials because the last visceral arch lacks an interbranchial septum and its associated muscles. Another muscle associated with each interbranchial septa is the **branchial adductor**, which can be viewed by cutting the dissected septum and its branchial arch. Each of the five branchial adductors lies medial to their corresponding branchial arches.

LEVATORS

The levators are located on the dorsal side above the spiracle. Their contraction raises the visceral arches. The levator series consists of three muscles on each side of the body: the **levator palatoquadrati**, the **levator hyoideus**, and the **cucullaris**. Identify these muscles as shown in Figures 3.1 and 3.5:

Levator palatoquadrati: a small muscle immediately anterior to the craniomaxillaris, it is the levator of the first visceral arch. Its origin is on the otic capsule, and its insertion is on the palatoquadrate cartilage near the quadratomandibularis. Its action is to raise the upper jaw.

Levator hyoideus: a band of fibers immediately deep to the epihyoideus. It extends from its origin on the otic capsule to its insertion on the hyomandibularis. It is the levator of the second visceral arch; as such, its action is to raise the hyoid (second visceral) arch. Do not attempt to dissect it as it is closely invested with the epihyoideus.

Cucullaris: a long triangular muscle dorsal to the dorsal superficial constrictors. It extends from its origin on fascia of the dorsal longitudinal bundles to its insertion on the scapular process of the pectoral girdle. Some of its fibers also insert on the epibranchial cartilage of the last visceral arch. As the levator of the remaining visceral arches, its action is their elevation and that of the pectoral girdle.

INTERARCUALS

The interarcuals are a series of small muscles that lie deep to the cucullaris and the dorsal longitudinal bundles. They extend between adjacent pharyngobranchial cartilages or between pharyngobranchials and their corresponding epibranchials. Their contraction draws the branchial arches together, thereby expanding the pharynx.

To view the interarcuals, the cucullaris and the dorsal longitudinal bundles must be spread apart. Do this with blunt probes or your fingers, and do not cut any fibers. The space exposed is called the **anterior cardinal sinus**; the interarcuals are located within.

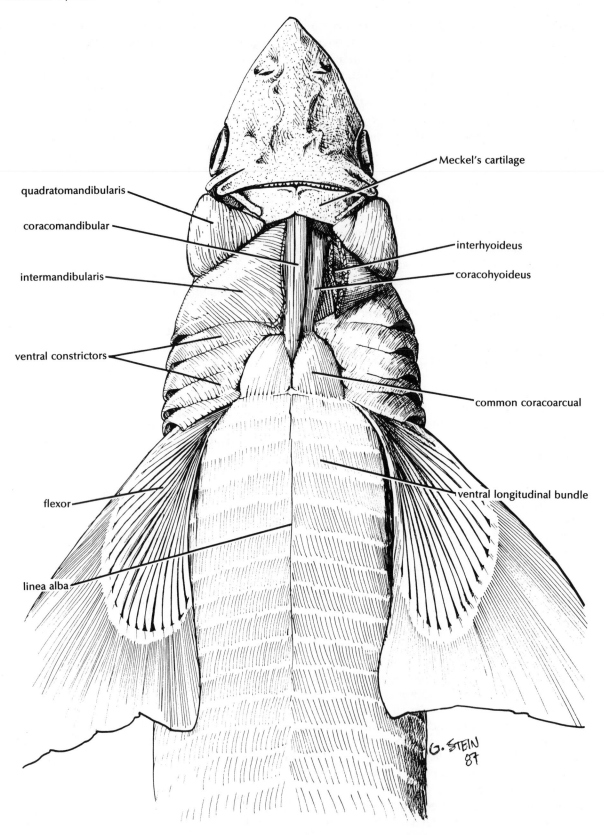

quadratomandibularis

coracomandibular

intermandibularis

ventral constrictors

flexor

linea alba

Meckel's cartilage

interhyoideus

coracohyoideus

common coracoarcual

ventral longitudinal bundle

G. STEIN 87

FIGURE 3.4.   Muscles of the head and cranial trunk, ventral view. The intermandibularis and interhyoideus muscles on the left side are shown cut and reflected.

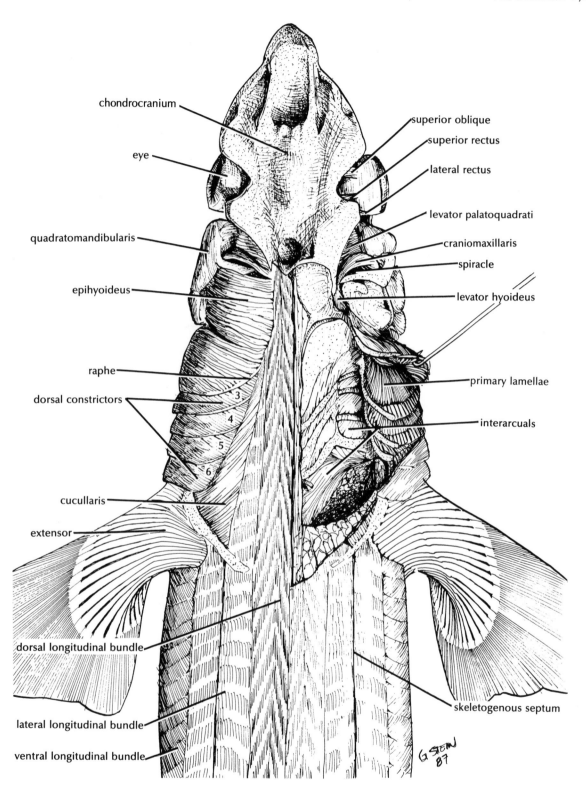

**FIGURE 3.5.   Muscles of the head and cranial trunk, dorsal view.
Deeper muscles are shown on the right side.**

HYPOBRANCHIAL MUSCLES

The hypobranchial muscles are located below the gills, as their name implies. They are composed of somatic muscle tissue and are collectively considered as an anterior extension of the hypaxial musculature. They serve to strengthen and to elevate the floor of the mouth cavity (open the mouth) and play a role in expansion of the branchial pouches. With the aid of Figure 3.4, identify the following hypobranchial muscles on the ventral side of your specimen:

Common coracoarcual: triangular muscles that originate from the coracoid bar. Their fibers pass anteriorly to insert onto fascia that also serve as a point of origin for the coracomandibular and coracohyoid muscles (below).

Coracomandibular: paired muscles that extend from their origin on fascia from the common coracoarcuals anteriorly to insert on the posterior edge of Meckel's cartilage. They are located deep to the intermandibularis and interhyoideus muscles. On the same side on which you previously cut the intermandibularis to expose the interhyoideus, cut the interhyoideus, as shown in Figure 3.4. The coracomandibular is the more medial slender muscle that is now exposed.

Coracohyoideus: paired muscles located lateral and dorsal to the coracomandibular. Their origin is on the fascia of the common coracoarcuals, and they extend anteriorly to insert on the basihyal cartilage of the hyoid arch. Note the dark-colored mass between the anterior portion of the coracomandibulars and the coracohyoids. This is the **thyroid gland** of the endocrine system.

Coracobranchial: located deep to the coracohyoideus, they consist of five segments that fan out to insert on branchial arches two through five and the basibranchial cartilage. Due to their deep position, they will be observed at a later stage in the dissection.

Internal Anatomy

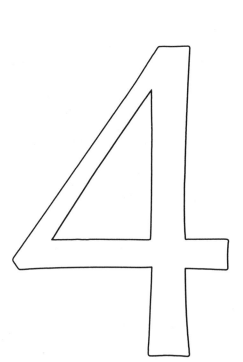

THE DOGFISH REPRESENTS a general internal organization that is common to fishes and other vertebrates. Its most conspicuous internal feature, like that of other vertebrates, is a large central cavity called the **coelom**. The coelom is an epithelium-lined space that contains the visceral organs and a small amount of fluid. As in all fish, it is divided into two parts: a small cranial portion that surrounds the heart, called the **pericardial cavity**, and the remainder called the **pleuroperitoneal cavity**. The two cavities are separated by a fold called the **transverse septum**. In dogfish the two cavities may communicate by a small opening through the transverse septum called the **pericardioperitoneal canal**.

In more advanced vertebrates that are air-breathing, the pleuroperitoneal cavity has come to be divided into two major portions. The more cranial portion consists of a pair of cavities that are lateral to the heart and surround the lungs (the **pleural cavities**). The caudal portion contains the visceral organs (the **peritoneal cavity**). In mammals, these two portions are completely separated by the muscular diaphragm. The diaphragm is in part embryonically derived from the transverse septum, providing us with a clue of the evolutionary origin of this more advanced structure.

In this chapter, the internal anatomy of the dogfish is presented as it appears when you first expose the coelom before manipulation of any body structures. This, therefore, represents an overview of the locations of the major internal organs and membranes of the dogfish. In subsequent chapters, their categorization into body systems and their functional roles will be discussed.

DISSECTION OF THE BODY WALL

In order to expose the coelom, the body wall must first be dissected. To do this, follow the procedure below.

1. Place your specimen on a tray on its back to expose the ventral side. Using sharp scissors, make a shallow incision through the skin layer from a point immediately cranial to the cloaca to the pectoral girdle slightly to the left of the midventral line (Fig. 4.1). It is important to avoid cutting along the midline near the pectoral girdle

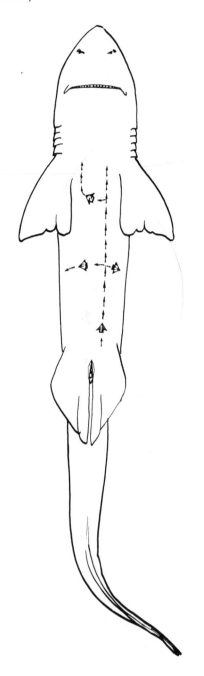

FIGURE 4.1.   Dissection diagram of the dogfish, ventral view

vic girdle. Extend the cut on both sides to the lateral line. Note the body layers that have been cut. From the surface passing deep, they are the **skin**, a thin layer of **connective tissue**, the **hypaxial musculature**, and the smooth membrane lining the pleuroperitoneal cavity, called the **parietal peritoneum**. Now pull back the flaps completely to observe the coelom and its contents.

COELOMIC MEMBRANES

Closely inspect the internal anatomy of your dogfish. Notice the large central cavity that contains the visceral organs. This is the **pleuroperitoneal cavity**. At its cranial extremity is the membranous **transverse septum**, which partially separates it from the **pericardial cavity** surrounding the heart. The coelomic membranes are associated with these cavities in that they line their inner walls and cover the surfaces of organs that are contained within. The membranes associated with the pericardial cavity are collectively called **pericardium**, and those associated with the pleuroperitoneal cavity are called **peritoneum**.

PERICARDIUM

The pericardium surrounds and protects the heart (Figs. 4.2 and 5.4). It consists of two distinct layers: an outer **parietal pericardium**, which lines the inner wall of the pericardial cavity, and an inner **visceral pericardium**, which closely attaches to the outer surface of the heart.

PERITONEUM

The peritoneum is an extensive membrane that lines the pleuroperitoneal cavity and surrounds the visceral organs. The portion that lines the internal surface of the body wall is called the **parietal peritoneum**, and that which covers most of the visceral organs is the **visceral peritoneum**.

In addition to the peritoneal membranes, there are **peritoneal folds**. These extensions of the peritoneum help suspend the visceral organs within the pleuroperitoneal cavity. Locate the following peritoneal folds in the dogfish (Figs. 4.2 and 5.4):

Dorsal mesentery: a double layer of peritoneum that anchors visceral organs to the dorsal body wall. It consists of the following:

    Mesogaster: attaches to the stomach.

    Mesointestine: also called the **mesentery proper**. It attaches to the intestine.

    Mesorectum: attaches to the rectum or caudal portion of the large intestine.

Ventral mesentery: a reduced portion of the peritoneum that persists in adults as several specialized ligaments. These provide support for some visceral organs to the ventral body wall and include the following:

in order to prevent damage to underlying structures. Now cut through the muscle layer by making a deeper incision along the same path as the first incision through the skin.

2. Carefully extend the incision made above from the coracoid bar of the pectoral girdle to the level of the second gill slit. Cut through the body wall, but do not cut deeply.

3. Make a transverse incision through the body wall about midway between the pectoral girdle and the pel-

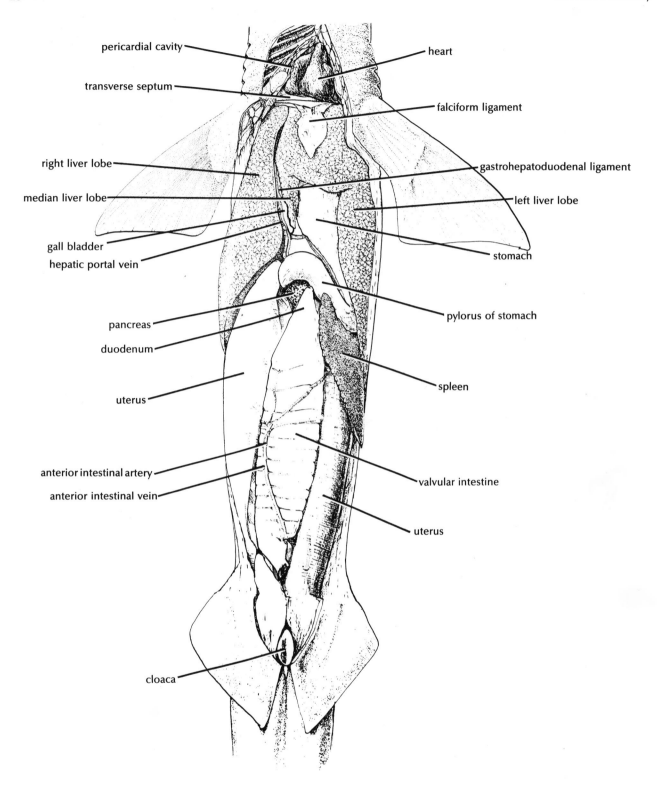

FIGURE 4.2.   Internal organs of the coelom, ventral view of a female specimen

Falciform ligament: attaches to the liver near its cranial end along the midventral line.

Gastrohepatoduodenal ligament: a slender membrane that passes from the liver to the stomach and duodenum. At its caudal end, it divides into two parts—a right **hepatoduodenal ligament** and on the left the **gastrohepatic ligament**.

Lateral mesentery: peritoneal folds that anchor the gonads to the lateral and dorsal body walls. They include the following:

Mesovarium: supports the ovaries of the female.

Mesorchium: supports the testes of the male.

MAJOR ORGANS OF THE COELOM

Identify the following organs located within the coelom, using as little manipulation as possible (Fig. 4.2):

Heart: the single organ of the pericardial cavity.

Liver: the large yellowish structure at the cranial end of the pleuroperitoneal cavity. Its two major lobes, right and left, extend nearly to the pelvic girdle.

Stomach: a large smooth-textured structure between the right and left liver lobes.

Spleen: a dark-colored organ near the caudal end of the stomach. It is a component of the lymphatic system.

Intestine: a caudal continuation of the stomach. Continuing caudally, it consists of three sections: the **duodenum**, the **valvular intestine**, and the **colon**.

Pancreas: a white multilobed organ that wraps around the duodenum.

Gallbladder: a small greenish sac near the midline of the liver.

Kidney: paired brown structures lying between the dorsal body wall and the parietal peritoneum on either side of the vertebral column.

Gonads: paired oval structures deep to the stomach. To view them, push the stomach carefully to one side. If your specimen is a female, the gonads are called **ovaries**. If your specimen is a male, they are **testes**.

The Digestive & Respiratory Systems

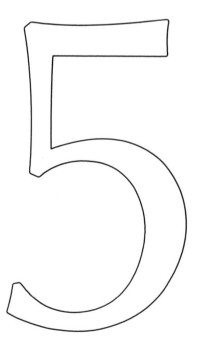

THE DIGESTIVE AND RESPIRATORY systems are combined in this chapter because of the close location of some of their components. This organizational strategy will permit you to examine the close structural relationships that exist between these two organ systems in the region of the pharynx.

The digestive system of the dogfish consists of a long continuous tube that extends through the pleuroperitoneal cavity from the mouth to the cloacal aperture and a number of closely associated accessory structures. The central tube, also known as the **gastrointestinal (G.I.) tract** or **alimentary canal**, is the site where the functions of mechanical digestion, chemical digestion, nutrient absorption, and storage and elimination of solid waste material take place. Each of these processes occurs within a compartment of the tract that is specialized to accommodate it. The specialized compartments, or organs, of the G.I. tract are the mouth, pharynx, esophagus, stomach, intestine, and cloaca.

The accessory structures of the digestive system are closely associated with the G.I. tract in that they are either located within it or communicate with it by means of a duct. They include the teeth, tongue, liver, gallbladder, and pancreas.

The primary functions of the respiratory system are to provide the body with a continuous supply of oxygen and to remove the metabolic waste product, carbon dioxide. In fish, this is accomplished by the diffusion of these gases between the water and specialized structures in the walls of the pharynx called **gills**. Efficiency in this process is increased by the enlargement of the potential surface area within the gills through gill filaments and lamellae.

ORAL AND PHARYNGEAL REGIONS

The oral and pharyngeal regions of the dogfish contain structures that are digestive or respiratory in function or both. In the following dissection procedure, you will first examine the structures of the oral cavity and pharynx that play a functional role in both systems. A study of the respiratory structures in the wall of the pharynx will follow.

ORAL CAVITY

The **mouth** or **oral cavity** is the space between the external ridge of the teeth and the internal openings of the spiracles. To view its features, cut through the left corner of the mouth. Continue this cut through the Meckel's cartilage, cartilages of the branchial arches, and the hyoid arch, as shown in Figure 5.1. Identify the following features of the oral cavity:

Teeth: Present in rows on the upper and lower jaw rims, they are similar in shape to each other (**homodont**). They are homologous with the placoid scales, as both are derived from the dermis.

Tongue: forms the floor of the oral cavity. It is not a true tongue as it is not capable of movement, but it is homologous to the functional tongue of higher vertebrates.

Spiracle: Observe the internal openings of the two spiracles in the roof of the mouth. Recall that they represent the first gill slits and function as a passageway for water when the mouth is closed.

PHARYNX

The pharynx is a muscular chamber that extends as a posterior continuation of the oral cavity to the esophagus. As a digestive organ, it provides passage for food to the esophagus. As a respiratory organ, it allows water to flow from the mouth and spiracles to the gill slits, which occupy its lateral walls. Identify the five pairs of **internal gill slits** in your specimen, and locate the following features of the gills (Figs. 5.1, 5.2, and 5.3):

Branchial arch: cartilaginous structures that provide support to the gills. An individual branchial arch corresponds to each of the 5 gill pairs.

Gill raker: fingerlike projections from the branchial arches that ring the internal gill slits and serve as strainers.

Branchial pouch: a small chamber that is entered from the internal gill slit. It opens to the exterior via an **external gill slit**.

Interbranchial septum: a connective tissue bar that separates adjacent branchial pouches and slits. Note the external portion of an interbranchial septum. Its flaplike projection serves as a valve for opening and closing the external gill slits. Note also that each septum is attached to a branchial arch on its medial side.

Gill filaments: the actual respiratory portion of the gills. The numerous filaments bear platelike folds called **primary lamellae**, which increase their surface area. The lamellae are attached to the surfaces of the septa and contain the branchial blood vessels and capillaries. An interbranchial septum with filaments and lamellae on both sides constitutes a complete gill, or **holobranch**. A septum with filaments and lamellae on one side, such as in the first gill, is termed a **hemibranch**.

Secondary lamellae: Visible only with a hand lens, the minute secondary lamellae cover the surface of each primary lamella. They serve to increase further the surface area of the gill.

Pseudobranch: Cut open the spiracle on the left side and notice the small hemibranch present on the spiracular valve. The lamellae here are thought to be rudimentary in function; hence the name, *pseudobranch*.

Branchial muscles: muscles that constitute the actual bulk of the gill. They include the **adductor muscle** that lies medial to each gill arch and an **interbranchial muscle** and a **superficial (septal) constrictor muscle** that extend into the septum.

Branchial vessels: With a hand lens, locate the small blood vessels within a single gill. The **afferent branchial artery** may be viewed near the middle of the septum lateral to the gill arch, and the **efferent branchial artery** is at the base of the primary lamellae on each side. The afferent artery may not be injected with colored latex, as it carries unoxygenated blood from the heart to the gills.

DIGESTIVE ORGANS OF THE PLEUROPERITONEAL CAVITY

The remainder of this chapter will guide you through dissection of the G.I. tract and the digestive accessory organs that are located within the pleuroperitoneal cavity. With your specimen on its back to expose the ventral side, identify the following digestive organs and their features (Figs. 5.4 and 5.5):

LIVER

The liver is composed of three lobes: two very large **lateral lobes** (right and left) and a central **median lobe**. It is attached to the transverse septum by a narrow **coronary ligament** and to the ventral body wall by the **falciform ligament**. The primary functions of the liver include production of bile, interconversions of storage products, detoxification or storage of certain wastes, and storage of fats (which produce buoyancy).

Exiting in a caudal direction from the liver and located parallel to blood vessels in the region is the **bile duct**. This tube transports bile from the liver to the gallbladder (via the **cystic duct**) for storage or to the duodenum.

GALLBLADDER

The gallbladder is a thin-walled sac that stores bile from the liver, which it receives via the cystic duct. It is a greenish elongate structure visible at its attachment to the median lobe of the liver.

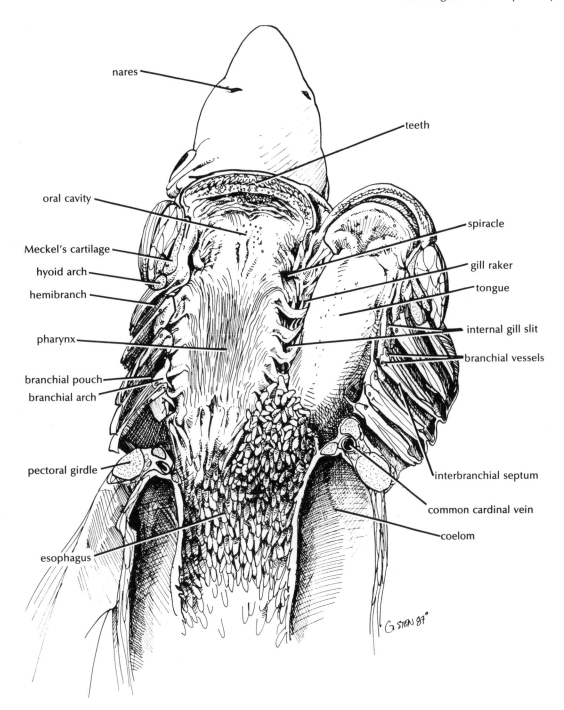

FIGURE 5.1. The oral cavity and pharynx, ventral view. The jaw is cut on the right side and reflected laterally.

ESOPHAGUS

The esophagus is a very short constriction between the pharynx and the stomach. Push the left lateral lobe of the liver aside, and you will notice that it appears as the cranial portion of the stomach. There is no external distinguishing line of division between the two organs.

STOMACH

With the lateral liver lobes pushed aside, locate the long slender stomach. Near its caudal end it narrows before continuing cranially a short distance, giving it a J-shaped appearance. Its medial side is called the **lesser curvature**, and its lateral side is the **greater curvature**. It

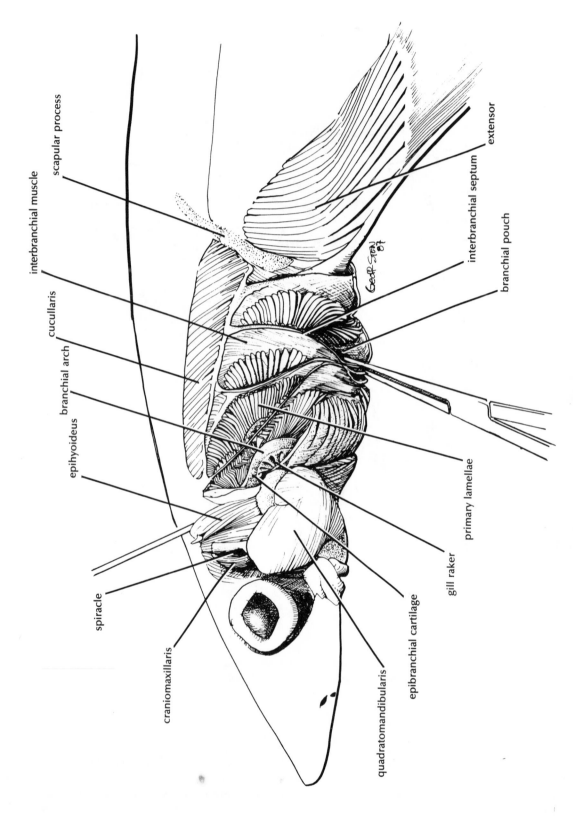

FIGURE 5.2. The gills and associated structures, lateral view. The superficial constrictor muscles have been removed to reveal the underlying gills.

interbranchial muscle

scapular process

cucullaris

branchial arch

epihyoideus

spiracle

craniomaxillaris

quadratomandibularis

epibranchial cartilage

gill raker

primary lamellae

branchial pouch

interbranchial septum

extensor

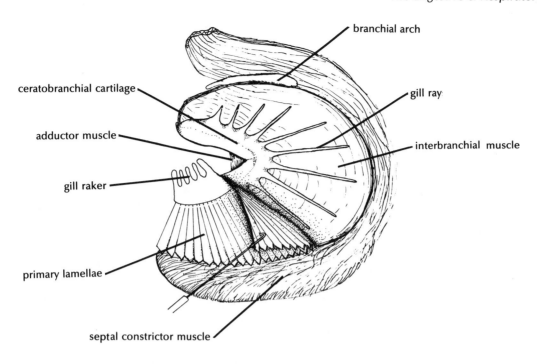

FIGURE 5.3. A single interbranchial septum with its associated gill, muscle, and skeletal components. The gill filaments are shown partially pulled outward to reveal deeper structures.

may be divided into three segments, which are, from cranial to caudal, the **cardia**, the **body**, and the **pylorus**. It terminates at an internal valve that opens to the duodenum called the **pyloric valve**.

At the cranial end of the **pleuroperitoneal cavity**, cut through the median lobe of the liver and remove it. Now make a median incision through the ventral wall of the esophagus and extend this cut halfway down the stomach. Examine the internal surface of the esophagus and stomach. The **papillae**, or fingerlike projections, identify the internal surface of the esophagus. The wrinkles present on the internal surface of the stomach are called **rugae**, and the inner epithelial lining is the **mucosa**. External to the mucosa, the stomach wall is composed of connective tissue and muscle. Recall that the stomach is anchored to the dorsal body wall by the **mesogaster**.

INTESTINE

The intestine is a continuation of the G.I. tract from the pyloric valve. Recall that its cranial segment is attached to the dorsal body wall by the **mesointestine** or **mesentery proper**, and its caudal segment is attached by the **mesorectum**. It is functionally important in chemical digestion and nutrient absorption.

The three segments of the dogfish intestine are, from cranial to caudal, the **duodenum**, the **valvular intestine**, and the **colon**. The duodenum is a short segment that extends to the right from its union with the stomach before continuing caudally. It receives bile from the bile duct. The valvular intestine appears as an expanded compartment that continues caudally from the duodenum. Cut open this structure by making a shallow incision along one side between the large blood vessels that travel lengthwise to its wall. The **spiral valve** located within serves to increase the absorptive surface area, which is the same effect that is maximized in mammals by the extensive length of the small intestine. The caudal segment, the colon, terminates at the **anus**, which opens into the cloaca. Near the colon is a small narrow structure called the **rectal** or **digitiform gland**. It communicates with the colon via a duct and is thought to play a role in salt excretion.

PANCREAS

The pancreas is a lightly colored structure located at the curve of the duodenum. It consists of an oval **ventral lobe** that adheres to the ventral surface of the duodenum and a larger **dorsal lobe** that extends from the spleen to the dorsal side of the duodenum. The two lobes are connected by a narrow band called the **isthmus**. The pancreas functions in the production of digestive enzymes that are collectively called **pancreatic juice**, which it re-

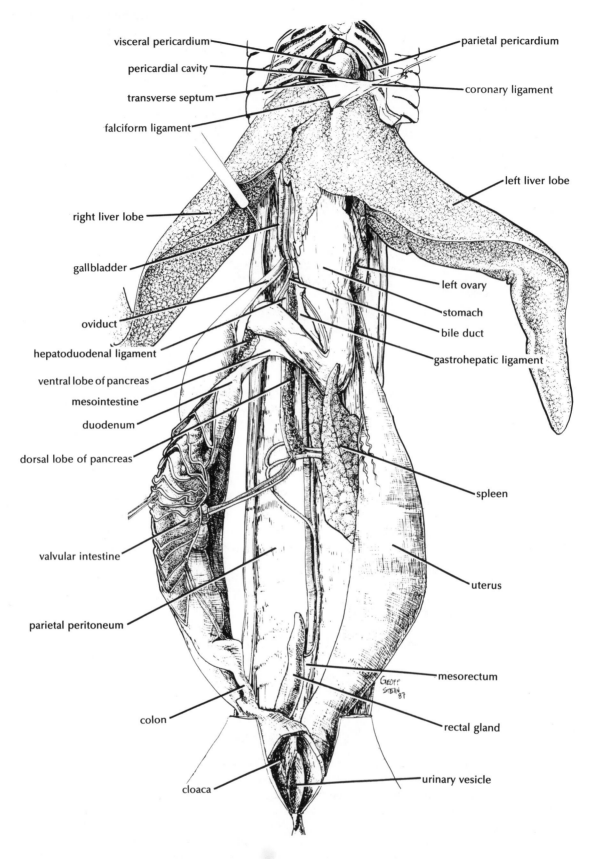

visceral pericardium

parietal pericardium

pericardial cavity

transverse septum

coronary ligament

falciform ligament

left liver lobe

right liver lobe

gallbladder

left ovary

stomach

oviduct

bile duct

hepatoduodenal ligament

gastrohepatic ligament

ventral lobe of pancreas

mesointestine

duodenum

dorsal lobe of pancreas

spleen

valvular intestine

uterus

parietal peritoneum

mesorectum

colon

rectal gland

cloaca

urinary vesicle

GEOFF
STEIN
'87

**FIGURE 5.4.   Digestive organs and associated structures of the coelom, ventral view
of a female specimen. The liver lobes are spread laterally and the intestine is pushed to
the right lateral side. The valvular intestine has been cut to reveal the internal spiral valve.**

30

leases into the duodenum. It also is an endocrine gland, in that it secretes the hormones **insulin** and **glucagon** into the bloodstream.

CLOACA

The cloaca is a chamber that receives solid waste material from the G.I. tract through the anus. It eliminates this waste to the exterior through the **cloacal aperture**. As in all vertebrates but teleost fishes and placental mammals, the cloaca also receives liquid waste from the urinary bladder and sex cells from the gonads.

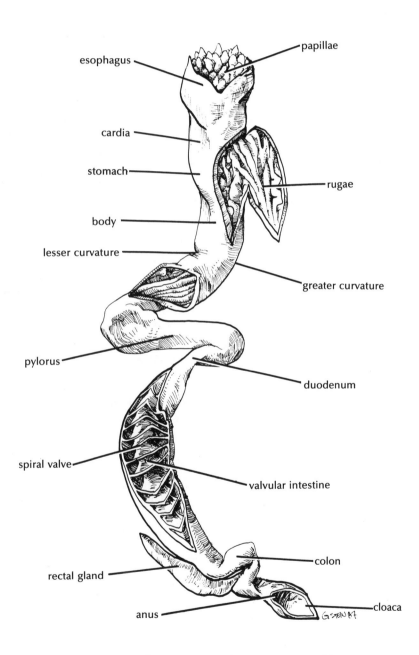

esophagus

papillae

cardia

stomach

rugae

body

lesser curvature

greater curvature

pylorus

duodenum

spiral valve

valvular intestine

rectal gland

colon

anus

cloaca

FIGURE 5.5.   The G.I. tract, as it appears removed from the coelom and partially sectioned

The Circulatory System

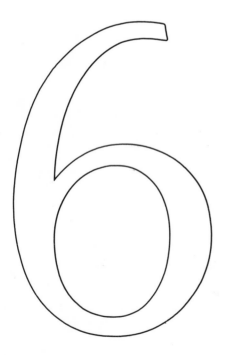

THE PRIMARY FUNCTION of the circulatory system is to serve as a transport vehicle for a number of substances, including oxygen, carbon dioxide, nutrients, nitrogenous wastes, and hormones. It consists of a **cardiovascular** component (the **heart**, **arteries**, **blood capillaries**, and **veins**), a **lymphatic** component (the **spleen**, **lymphatic vessels**, and **lymphatic capillaries**), and a liquid medium within which substances are suspended (**blood** and **lymph**). As a whole, the circulatory system is an extremely extensive network that invests every organ and most tissues. It is, therefore, necessary to limit the following dissection protocol to a study of the heart, the major arteries and veins, and their primary tributaries.

The circulatory system of the dogfish provides a good example of a generalized arrangement similar to that of most fishes. This arrangement is very primitive when compared to those of higher vertebrates. It is characterized by the presence of a heart with a single atrium and ventricle, an important circulatory route that passes through the gills for oxygenation of the blood, and a very low vascular pressure with sluggish circulation. To help reduce resistance to blood flow in regions where the pressure is extremely low, veins have expanded to contain large internal chambers. These enlarged veins are called **sinuses**. Of particular interest in the dogfish is the high osmolarity of the blood, which is primarily due to the high levels of the metabolic waste product, urea. If mammalian blood were to exhibit such concentrations, it would result in a severely toxic condition called *uremia*. Presumably, this condition has enabled sharks to survive in salt water and in other aquatic environments that are restrictably hostile to other fish.

EXTERNAL FEATURES OF THE HEART

With your specimen on its back to expose the ventral side, examine once more the pericardial cavity and the heart within. The dogfish heart is an S-shaped tube with its four chambers in linear sequence. They are, from caudal to cranial, the **sinus venosus**, **atrium**, **ventricle**, and **conus arteriosus**. To observe the sinus venosus and atrium on the dorsal side, lift the caudal end of the ventri-

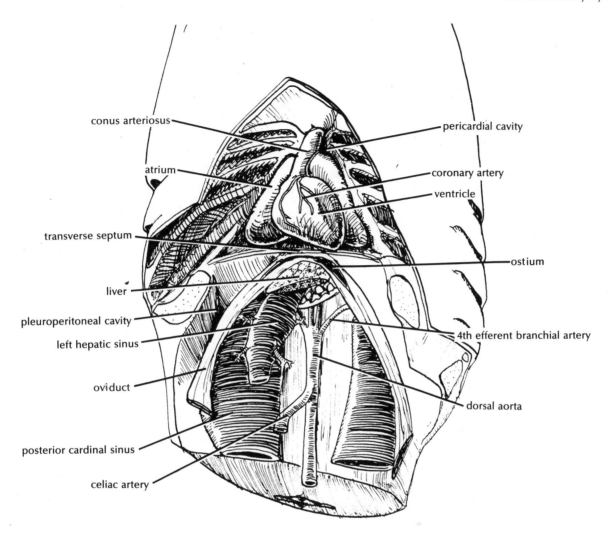

FIGURE 6.1.   The heart and associated structures, ventral view. The sinus venosus is not visible since it lies dorsal to the ventricle.

cle. Identify the following external features of the heart (Fig. 6.1). The heart chambers will be dissected later following examination of the blood vessels.

Ventricle: the thick-walled oval portion of the heart that occupies the caudoventral portion of the pericardial cavity. It receives blood from the atrium, and its contraction pushes blood into the conus arteriosus.

Conus arteriosus: a tubelike cranial extension of the ventricle that continues to the cranial end of the pericardial cavity. Its contraction helps propel blood to the gills for oxygenation.

Atrium: With the caudal end of the ventricle lifted, the atrium can be viewed on the dorsal side. It is a thin-walled, bilobed chamber that receives blood from the sinus venosus and passes it to the ventricle.

Sinus venosus: a triangular chamber caudal to the atrium. It is also thin walled and extends anteriorly from the transverse septum to lie between the two lobes of the atrium. It cannot be observed from the ventral side because the ventricle, to which it lies dorsal, obscures it. The sinus venosus receives blood from the body via veins and sinuses.

THE VENOUS SYSTEM

You are now ready to begin study of the major blood vessels of the dogfish. Most preserved specimens have had colored material (usually latex) injected into some of their vessels after death. This color coding distinguishes between arteries and veins and makes the blood vessels tougher and more elastic. Doubly injected specimens have a red material in their arteries and a blue material in their hepatic portal vessels. In triply injected specimens, usually a yellow material is injected into the systemic veins. As you dissect the blood vessels, keep in mind that they are subject to considerable variation and may therefore be a bit different in individual specimens. This is particularly true of the venous system.

Veins are blood vessels that transport blood toward the heart. In the dogfish and other fish, the venous system consists of two major **portal systems** and a series of **systemic veins**. A portal system consists of vessels that transport blood from one capillary network to another in a direction leading toward the heart. Systemic veins carry blood directly toward the heart. Using Figures 6.2 through 6.5 as guides, identify the components of the venous system, as indicated below.

HEPATIC PORTAL SYSTEM

The hepatic portal system is a series of vessels that transport blood from the G.I. tract to the liver. Following a meal, this blood is laden with newly absorbed nutrients that are to be processed by the liver. With the G.I. tract intact, locate the following vessels (Figs. 6.2 and 6.3):

Hepatic portal vein: the large vein located parallel to the bile duct in the gastrohepatoduodenal ligament (Fig. 6.2). It is the principle vein of the hepatic portal system and is formed near the cranial end of the dorsal lobe of the pancreas by the convergence of the following three veins:

    Gastric vein: originates from the central part of the stomach.

    Lienomesenteric vein: originates near the spleen where it is formed by the union of the posterior intestinal vein from the spiral intestine and the posterior lienogastric vein from the spleen and stomach. From this origin it passes cranially along the surface of the dorsal lobe of the pancreas until it merges with the hepatic portal vein. Along its length, it receives numerous **pancreatic veins**, which drain the pancreas.

    Pancreaticomesenteric vein: originates as a cranial continuation of the **anterior intestinal vein** at a point immediately deep to the ventral lobe of the pancreas, which drains the valvular intestine. The pancreaticomesenteric also receives the **pyloric vein** from the pylorus and the **anterior lienogastric vein** from the spleen.

Trace the entire length of the hepatic portal vein to its convergence with the liver. Cut through a segment of the liver and note that, once within, the hepatic portal vein branches extensively. Eventually, these branches will carry blood into microscopic chambers called **sinusoids**,

which later drain into two large **hepatic veins**. The hepatic veins exit the liver and return blood to the heart for recirculation.

RENAL PORTAL SYSTEM

The renal portal system shunts blood from the tail to the capillary networks within the kidneys for filtration. Locate the following components of this portal system in your specimen (Fig. 6.4):

Caudal vein: Make a cross section across the tail just caudal to the cloacal aperture, and note the **caudal vein** and **artery** within the hemal arch. The caudal artery is dorsal to the vein and is usually injected, whereas the vein is not. The unpaired caudal vein is the origin of the renal portal system.

Renal portal vein (paired): The caudal vein divides, or *bifurcates*, at a point cranial to the pelvic girdle into the paired renal portal veins. The renal portals continue cranially along the dorsal surface of each kidney and usually contain red latex (Fig. 6.4). Blood from the renal portal vein enters the kidney via small branches called **afferent renal veins**, which branch further within the kidney to form capillaries. Blood is collected via **efferent renal veins**, which enter the **posterior cardinal veins** (below) for the return trip to the heart.

SYSTEMIC VEINS

The systemic veins transport blood directly to the heart from tissues and organs. Locate the sinus venosus of the heart once again by lifting the base of the ventricle and examining the dorsal side of the heart. Make a transverse incision through the ventral wall of the sinus venosus and wash it out. Note the entrances of the hepatic veins, present near the center of the sinus. Also locate the larger openings at the caudolateral angles of the sinus. These are the entrances of the **common cardinal veins**. The sinus venosus receives blood from the liver via the hepatic veins and blood from the remainder of the body via the common cardinal veins. Identify the following veins and sinuses, using Figures 6.4 and 6.5 as guides:

Common cardinal vein (paired): Also called **ducts of Cuvier**, they unite with the sinus venosus at its lateral ends and drain the large **posterior cardinal sinus** on each side.

Posterior cardinal sinus (paired): To view these large chambers, push aside the liver and the caudal portion of the G.I. tract below the esophagus. The right and left posterior cardinal sinuses lie against the body wall dorsal to the esophagus. They transport blood from the **posterior cardinal veins** (below) to the common cardinal veins.

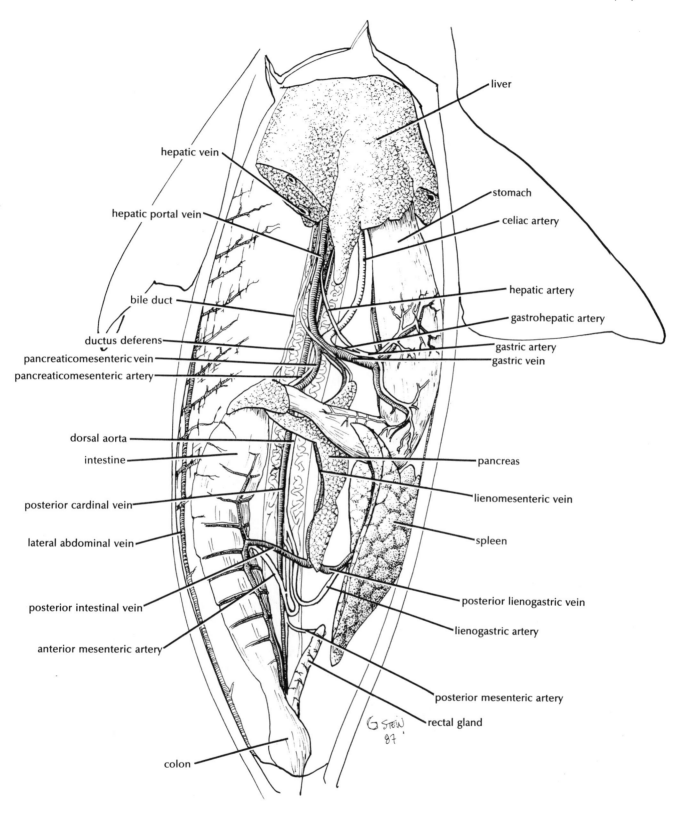

**FIGURE 6.2.   Coelomic blood vessels of a male, ventral view.
The right and left liver lobes are cut and partially removed.**

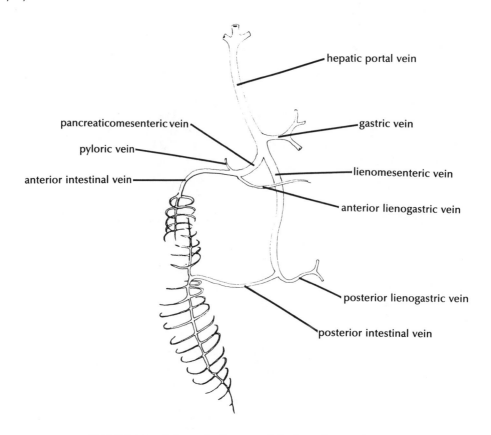

hepatic portal vein

gastric vein

pancreaticomesenteric vein

lienomesenteric vein

pyloric vein

anterior intestinal vein

anterior lienogastric vein

posterior lienogastric vein

posterior intestinal vein

FIGURE 6.3.   Schematic diagram of the hepatic portal system

Posterior cardinal vein (paired): originates as a cranial continuation of the efferent renal veins, which it drains. From here, each posterior cardinal vein extends cranially along one side of the dorsal midline to merge with the posterior cardinal sinus. Normally, the right posterior cardinal vein is longer and larger than the left.

Anterior cardinal sinus (paired): extends cranially from the common cardinal vein on each side at its union with the posterior cardinal sinus. The anterior cardinal sinus may be entered by making a shallow incision along the ventral margin of the cucullaris muscle above the external gill slits. The cavity deep to this muscle is the anterior cardinal sinus. Continue your dissection by removing the gill lamellae and superficial jaw muscles on one side. With a probe, explore the narrowed junction of the anterior cardinal sinus and posterior cardinal sinus and the cranial extension of the anterior cardinal sinus around the eyeball called the **orbital sinus**.

Inferior jugular vein (paired): narrow vessels that originate from the cranial surfaces of the posterior cardinal sinuses. They drain the floor of the branchial region. Each vein extends in a craniodorsal direction to the pericardial cavity. They communicate with the anterior cardinal sinus of each side by a narrow channel called the **hyoidean sinus**.

Subclavian vein (paired): a short vessel that unites with the common cardinal vein immediately lateral to the entrance of the inferior jugular vein on each side. The subclavian is formed by the union of the **subscapular vein**, the **brachial vein**, and the **lateral abdominal vein**.

Lateral abdominal vein (paired): originates from the pelvic girdle to drain the appendages and a portion of the trunk. Each passes cranially along the inside of the body wall to the subclavian.

Brachial vein (paired): extends from the caudal side of the pectoral fin, which it drains, to join the lateral abdominal vein to form the subclavian. Locate the brachial vein on the pectoral fin near its base, and trace it through the body wall to its union.

Subscapular vein (paired): originates near the dorsal midline and extends ventrally to unite with the subclavian at its point of union with the brachial. Do not trace it along its length.

THE ARTERIAL SYSTEM

The arterial system of the dogfish consists basically of two circulatory routes, both of which carry blood in a general direction away from the heart. The route that

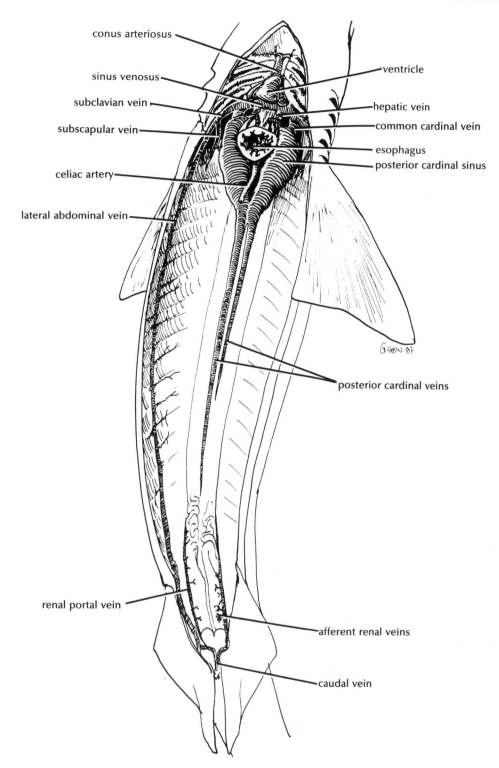

conus arteriosus

sinus venosus

subclavian vein

subscapular vein

celiac artery

lateral abdominal vein

ventricle

hepatic vein

common cardinal vein

esophagus

posterior cardinal sinus

posterior cardinal veins

renal portal vein

afferent renal veins

caudal vein

FIGURE 6.4.   Systemic and renal portal veins of the coelom, ventral view. All organs within the coelom are shown removed in this semidiagrammatic view.

exits directly from the heart is the **branchial circulation** and is composed of arteries that carry blood from the heart to the gills for oxygenation. The second route consists of a main vessel, the **dorsal aorta**, and its tributaries that transport newly oxygenated blood from the branchial region to its dispersion throughout body tissues.

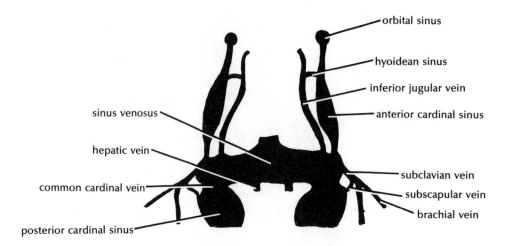

FIGURE 6.5.   Schematic diagram of the systemic veins of the head and cranial trunk

BRANCHIAL CIRCULATION

The branchial circulatory route brings oxygen-poor blood from the heart to the capillary network within the gills and oxygen-rich blood from the gills to the dorsal aorta. With your specimen on its back to expose the ventral side, identify the following arteries, as shown in Figures 6.6 and 6.7.

Ventral aorta: Locate the conus arteriosus of the heart once again, and note that it extends to the cranial end of the pericardial cavity. The ventral aorta begins as it emerges from the conus arteriosus out of the pericardial cavity. It is not injected. The ventral aorta contains two or three pairs of tributaries along its length before it bifurcates near the level of the hyoid arch, forming two vessels that extend dorsally. Each branch divides further into two arteries that lie on either side of the first branchial pouch. These divisions are the first two of the **afferent branchial artery** series, which consists of five pairs of arteries that supply the gill filaments on both sides of the corresponding bar. The caudal three pairs branch directly from the ventral aorta. Notice that each afferent branchial supplies the branchial pouch immediately caudal to it via smaller arteries that pass into the gill filaments and lamellae. It is here that the exchange of respiratory gases occurs.

Efferent collector loops (paired): a series of arteries that surround the five branchial pouches. They form a complete loop around the first four branchial pouches and an incomplete loop associated with the fifth branchial pouch. They carry freshly oxygenated blood away from the capillaries in the gill filaments. In order to trace the efferent collector loops,[1] swing the lower jaw that has been disarticulated on one side to expose the oral cavity and pharynx. If the jaw has not been cut, do this now by cutting through the angle of the jaw on the left side and continue caudally through the branchial arches, as shown in Figure 5.2, page 28. Carefully remove the mucus membrane from the roof of the mouth without damaging the underlying **efferent branchial arteries** and the **dorsal aorta**. This may be done by making a shallow incision along the base of the teeth and pulling the mucus membrane in a caudal direction with your fingers. Continue pulling until you reach the esophagus, then cut the membrane free and remove it. The mucus membrane covering the internal gill slits should also be removed by this procedure. Now remove the floor of the mouth in the same manner to expose the dorsal side of the heart and the **afferent branchial arteries**, which you should identify once more. After removing any remaining pieces of mucus membrane and connective tissue that may obscure the heart and vessels, your specimen should resemble Figure 6.6. With reference to this diagram, identify the following components of the efferent collector loops:

Posttrematic branch (paired): a branch of the collector loop that lies caudal to the branchial pouch. In each of the first four loops, it is the larger of the two vessels in diameter and is not present in the fifth branchial pouch.

Pretrematic branch (paired): the cranial portion of each collector loop. A pretrematic branch is present in each of the five branchial pouches on each side. Note that the posttrematic and pretrematic branches are continuous with each other to form a complete loop around the first four pouches, and adjacent collector loops are connected via small **intertrematic branches**.

External carotid artery (paired): a small artery that extends from its origin at the ventral end of the first collector loop to the lower jaw.

---

[1]Due to the level of difficulty in this procedure, your instructor may elect to provide a sample specimen that was prepared earlier. If this is the case, identify the branchial vessels from this specimen and Figures 6.6 and 6.7, and leave your specimen intact.

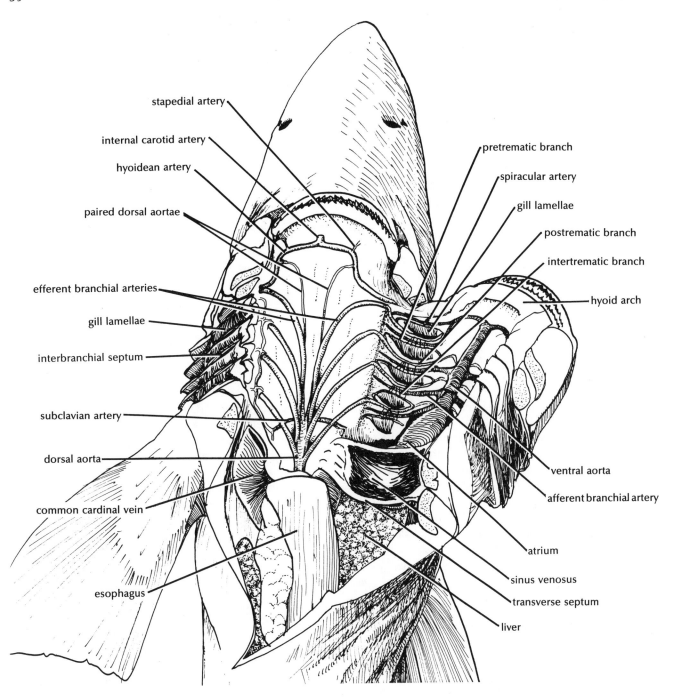

**FIGURE 6.6. Branchial arteries, ventral view. The jaw is cut on the right side
and reflected laterally to expose the oral cavity and pharynx,
and the mucus membrane is removed from the roof and floor of the mouth.**

Spiracular artery (paired): originates from the middle of
the pretrematic branch associated with the first branchial pouch. From here it passes dorsally to the spiracle
to supply the pseudobranch. On the cranial side of the
spiracle, it reappears to continue into the head region.

Hyoidean artery (paired): arises from the dorsal end of
the pretrematic branch corresponding to the first branchial pouch. It extends forward to the roof of the phar-

ynx to ultimately bifurcate into a lateral branch called
the **stapedial artery**, which passes to the orbit and
snout, and a medial branch called the **internal carotid
artery**, which unites with the internal carotid of the opposite side before entering the chondrocranium through
the carotid foramen as a single vessel to supply the
brain.

Commissural artery (paired): Also called the **hypobran-**

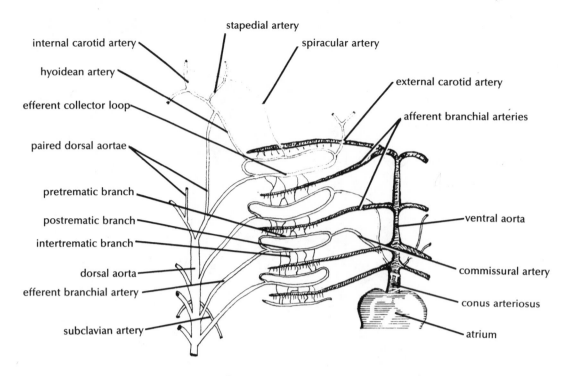

FIGURE 6.7. Schematic diagram of the branchial arteries

chial artery, it arises from the ventral end of usually the second and third efferent collector loops. From here it passes dorsally toward the heart to unite with the commissural artery of the opposite side on the dorsal side of the conus arteriosus. It then continues as a single vessel in a ventral direction until it joins the **coronary artery** of the heart ventricle to supply the musculature of the heart.

Efferent branchial arteries (paired): four pairs of vessels that continue from the posttrematic branches of the efferent loops caudally to the large, median dorsal aorta. Remove remaining connective tissue from these arteries on the left side, and trace them along their length.

Paired dorsal aortae: small vessels that extend from the hyoidean artery caudally to the first efferent branchial artery near its union with the single, large dorsal aorta.

THE DORSAL AORTA AND ITS BRANCHES

The dorsal aorta extends from its origin in the roof of the pharynx to the tail. Along its length, it gives off tributaries that supply all body structures caudal to the head with oxygenated blood. Identify the following tributaries of the dorsal aorta (Fig. 6.2):

Subclavian artery (paired): arises from the dorsal aorta between the second and third efferent branchial arteries. Separate the cranial portion of the esophagus from the body wall on the left side and trace the left subclavian. Note that it passes lateral to the posterior cardinal sinus, then continues ventrally to give off branches that supply the pectoral fin.

Brachial artery (paired): a major branch from the subclavian that passes into the pectoral fin. It parallels the brachial vein.

Ventrolateral artery (paired): a caudal branch of the subclavian that extends to the ventrolateral portion of the body wall. It may be viewed deep to the parietal peritoneum between the lateral abdominal vein and the midventral line.

Celiac artery: a large unpaired artery that supplies the viscera. From the dorsal aorta, it enters the pleuroperitoneal cavity and passes in a ventral and caudal direction along the right side of the stomach. Near its origin, the celiac artery gives off paired branches that pass to the gonads (**ovarian** or **testicular arteries**). At the dorsal lobe of the pancreas, it divides into the **pancreaticomesenteric** and **gastrohepatic arteries**.

Pancreaticomesenteric artery: parallels the pancreaticomesenteric vein along the dorsal side of the pylorus (of the stomach) to the intestine. At the intestine, it continues as the **anterior intestinal artery**, which parallels the vein of the same name.

Gastrohepatic artery: A short branch from the celiac that quickly divides into a **hepatic artery**, which parallels the hepatic portal vein to the liver, and a larger **gastric artery**, which parallels the gastric vein to the stomach.

Lienogastric artery: a branch from the dorsal aorta that may be viewed as it lies along the dorsal edge of the mesogaster. From here it passes to the spleen and the caudal part of the stomach.

Anterior mesenteric artery: arises from the dorsal aorta near the dorsal edge of the mesogaster just caudal to the

40

lienogastric artery origin. It parallels the posterior mesenteric vein and supplies the valvular intestine.

Posterior mesenteric artery: a ventral branch from the dorsal aorta that passes to the rectal gland and the caudal segment of the intestine.

Iliac artery (paired): Near its caudal end, the dorsal aorta submerges deep to the cloaca. Push aside the cloaca and locate the paired iliac arteries, which extend laterally in opposite directions before passing into the body wall. Continuing from the iliacs into each pelvic fin are the paired **femoral arteries**.

Caudal artery: the caudal continuation of the dorsal aorta into the tail. As observed earlier, it parallels the caudal vein, which is ventral to it.

INTERNAL STRUCTURES OF THE HEART

To examine the internal features of the dogfish heart, it must be removed from the pericardial cavity. To do this, cut the attachments of the sinus venosus to the transverse septum. Now cut through the ventral aorta near its union with the fourth and fifth afferent branchial arteries, and remove the heart. Cut across the walls of the heart along the left lateral side to expose the interior of the conus arteriosus, ventricle, and atrium, as shown in Figure 6.8. Using this diagram, locate the following internal features of the heart:

Sinus venosus: Locate this chamber, and recall that it was previously opened across the transverse plane. Note its thin walls, and identify the entrances of the hepatic sinus and common cardinal veins once again.

Atrium: Blood enters the atrium from the sinus venosus through a small opening called the **sinoatrial aperture**, which is bordered by a pair of lateral folds called the **sinoatrial valve**. This valve prevents backflow of blood into the sinus venosus from the atrium. Note the thin walls and large interior cavity that characterize the atrium. Despite its external bilobed appearance, the interior of the atrium is not divided.

Ventricle: From the atrium, blood is pushed into the ventricle through a small opening called the **atrioventricular aperture**. This opening is bordered by a pair of folds called the **atrioventricular valve**, which prevent backflow of blood into the atrium. Note the thickness of the cardiac muscle in the ventricular walls, the small internal cavity, and the muscular ridges of its internal walls.

Conus arteriosus: communicates directly with the ventricle, from which it receives blood. Backflow of blood into the ventricle is prevented by three rows of **semilunar valves** within the conus. Note its walls, which are thicker than those of the atrium and sinus venosus but thinner than those of the ventricle. From the conus arteriosus, blood is propelled to the ventral aorta and the branchial circulation.

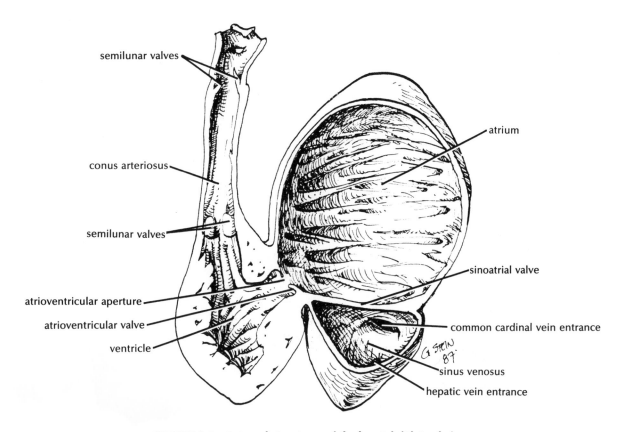

FIGURE 6.8. Internal structures of the heart, left lateral view

The Urogenital System

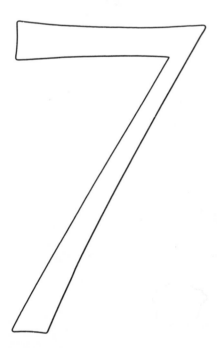

THE UROGENITAL SYSTEM has two primary functions that are quite different from each other: the formation and elimination of urine, and reproduction. These two functions are combined in one common system in fishes because of the common location and close association of a number of their organs. As you will discover, several urinary organs lie in direct contact with reproductive structures, and in some cases both functions take place within one organ. In mammals, urinary and reproductive functions are more anatomically separate, so these functions are often separated into two distinct systems.

The urinary functions within the urogenital system primarily involve the **kidneys**. These functions include the removal of nitrogen-containing waste products from the bloodstream, maintenance of the osmotic balance of fluids, and control of blood cell formation. Other structures that contribute to these functions include the rectal glands, the skin, and the gills.

A peculiarity of urinary function in cartilaginous fishes is the retention of urea in the bloodstream. Urea is retained by its reabsorption into the kidney tubules soon after it is initially removed and a reduction in the permeability of the gill membranes to urea to block its excretion. The presence of urea in the bloodstream results in an osmolarity greater than that in seawater, which causes water to enter the body through osmotic pressure. This is favorable to cartilaginous fishes because they require an inflow of water to compensate for the excessive removal performed by the primitive kidneys.

The reproductive functions of the urogenital system involve mainly the **gonads**, which produce the sex cells, or **gametes**. In the female, the reproductive organs are highly adapted for the production, storage, and fertilization of the female gametes (**ova**). In the spiny dogfish as in many other cartilaginous fishes, the female tract is modified for the internal retention of the young until embryonic development is complete. Most other fishes are egg-laying. In the male, the reproductive organs are adapted for the production of the male gametes (**spermatozoa**) and their release into the cloaca of the female.

In this chapter, dissection of the urogenital system is divided into the male and female systems. It is, therefore, advisable to combine dissection groups or switch specimens with another group to examine the urogenital system of both sexes.

THE MALE UROGENITAL SYSTEM

The male system in the dogfish is primarily located against the dorsal body wall. To view it clearly, the visceral organs must be removed. Do this now by cutting transversely across the liver about 5 cm from its attachment to the transverse septum. Also cut transversely through the esophagus at the same level, and remove the liver, G.I. tract organs, and associated structures. Be careful to avoid damaging the gonads, which lie dorsal to the liver and stomach. They should remain intact at this time. Now identify the following urogenital organs in the male specimen (Figs. 7.1 and 7.2).

KIDNEYS

The paired kidneys are long slender organs that lie against the dorsal body wall on either side of the midline. Based on their embryonic development and structural characteristics, they are categorized as **opisthonephroi**. This kidney type is distinguishable from the serially segmented kidney found among larval agnathans (**holonephroi**) and the caudal, compacted kidney present in reptiles, birds, and mammals (**metanephroi**).

To view the ventral surfaces of the kidneys, the parietal peritoneum must be removed, as they lie behind it. This positioning is called *retroperitoneal*. Lying on the ventral surface and extending along the length of each kidney is the tubular **ductus deferens**, which will be described below. In mature males, the tubules within the cranial two-thirds of the kidney have lost their urinary function, and instead provide a milky contribution to seminal fluid that passes through tiny ducts into the ductus deferens. This region of the male kidney is called **Leydig's gland**. At the cranial end of Leydig's gland, the tubules are further modified to serve as channels for the passage of spermatozoa en route from the testis to the ductus deferens. This small region is called the **epididymus**. The segment of the kidney that is active in urine production is the expanded caudal one-third.

TESTES

The paired testes, right and left, are located on either side of the dorsal midline adjacent to the cranial portion of the kidneys. They are attached to the dorsal body wall at the midline by the **mesorchium**. The testes are the male gonads; as such, they produce the male gametes (**spermatozoa**) and the primary male sex hormone (**testosterone**).

DUCTUS DEFERENS

The paired ductus deferens transport spermatozoa from the testes to the cloaca. They originate embryonically as **archinephric ducts** for the transport of urine from the kidneys to the cloaca. As the male matures, this function is lost and transferred to accessory ducts and replaced by reproductive functions. The urinary function of the archinephric ducts is retained in the female.

In a sexually immature male, each ductus deferens is a relatively straight tube. In a mature specimen, the ductus deferens is a tightly coiled tube that is conspicuous on the ventral side of the kidney. At its cranial end, it receives spermatozoa from the modified kidney tubules that constitute the epididymus. The epididymus receives spermatozoa from the testis via small **efferent ductules** that extend through the mesorchium. The efferent ductules cannot be viewed, however, due to their small size. As the ductus deferens extends caudally from its origin, it receives secretions from Leydig's gland until it nears its caudal end.

The caudal end of the ductus deferens is expanded to form a wide tube called the **seminal vesicle**. On the left side of your specimen, remove the seminal vesicle and observe a pair of small ducts that pass on either side of the dorsal midline. These are **accessory urinary ducts**. They transport urine from the kidneys and are present in males only. Both the seminal vesicles and the accessory urinary ducts unite caudally with the **sperm sac**.

SPERM SAC

The sperm sacs are a pair of small sacs located at the caudal end of the pleuroperitoneal cavity dorsal to the cloaca. They receive urine from the accessory urinary ducts and seminal fluid from the seminal vesicles. Carefully cut one open and notice the paired openings to these tubes. The small region caudal to their openings is the **urogenital sinus**. The urogenital sinus opens into the **cloaca** by way of a small projection called the **urogenital papilla**, which forms part of the dorsal wall of the cloaca.

CLOACA

The cloaca is a small chamber that receives material from the urinary tract, reproductive tract, and digestive tract. In the male dogfish, it receives urine and seminal fluid from the sperm sac via the urogenital papilla and solid waste from the rectum. Cut open the caudal end of the rectum that remains to view it.

SIPHON

The siphons are a pair of thin-walled sacs that lie outside the pleuroperitoneal cavity ventral to the pelvic fins in the male. They are located directly under the skin and open at their caudal end into the **clasper tube**, which

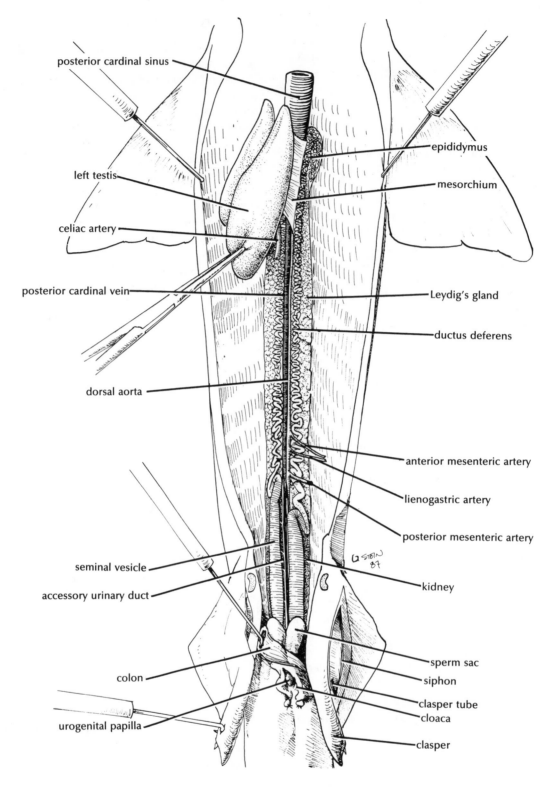

FIGURE 7.1. The urogenital system of the male dogfish, ventral view,
with digestive organs removed

passes along the dorsal side of the **clasper**. The siphons
secrete a lubricating fluid and a substance that contributes to the seminal fluid.

Fertilization in dogfishes, like all cartilaginous fishes,
is internal. During copulation, one of the claspers is in-
serted into the cloacal aperture of the female. Spermato-

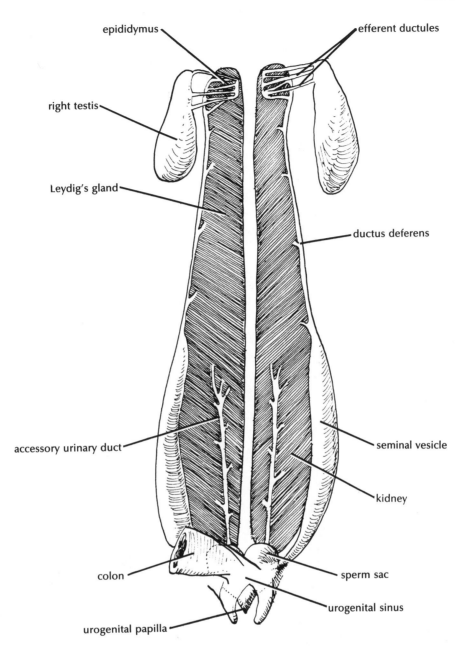

epididymus

efferent ductules

right testis

Leydig's gland

ductus deferens

accessory urinary duct

seminal vesicle

kidney

colon

sperm sac

urogenital sinus

urogenital papilla

FIGURE 7.2.   Schematic diagram of the male urogenital system, ventral view

zoa are released from the urogenital papilla into the cloaca. From the cloaca, it passes into the clasper tube and mixes with fluid from the siphon. This mixture then passes to the female through the clasper tube.

THE FEMALE UROGENITAL SYSTEM

Remove the visceral organs as described above for the male in the same manner, and identify the following organs of the female system (Figs. 7.3 and 7.4).

KIDNEYS

The female kidneys are similar in location and structure to those of the male with the exception of the cranial two-thirds, which is degenerate in the female due to a lack of reproductive involvement. As in the male kidney, the caudal one-third of each kidney is enlarged to accommodate urine production.

Urine is transported away from the kidney via the **Wolffian duct** (or **archinephric duct**) for there are no accessory urinary ducts as there are in the male. It is homologous to the ductus deferens of the male. The Wolffian

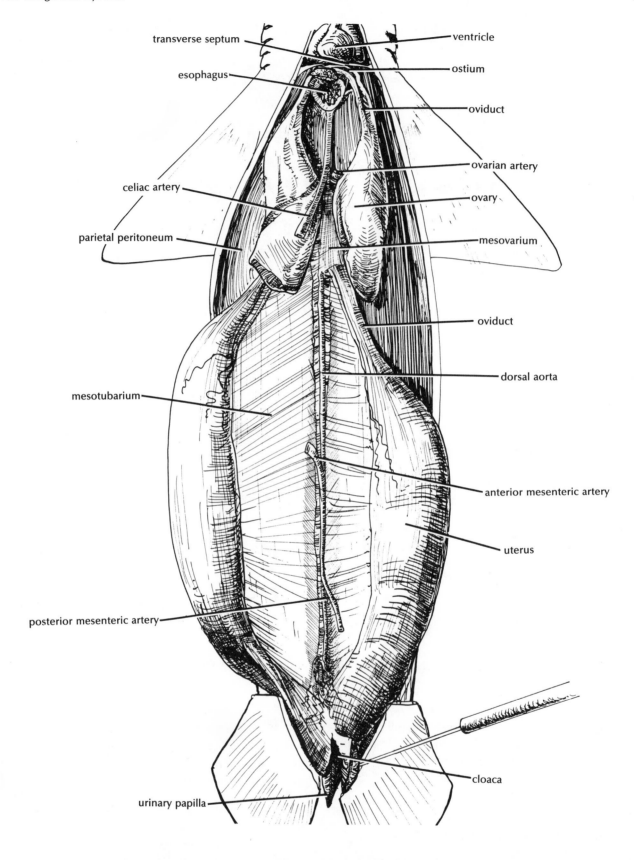

**FIGURE 7.3.   The urogenital system of the female dogfish, ventral view,
with digestive organs removed**

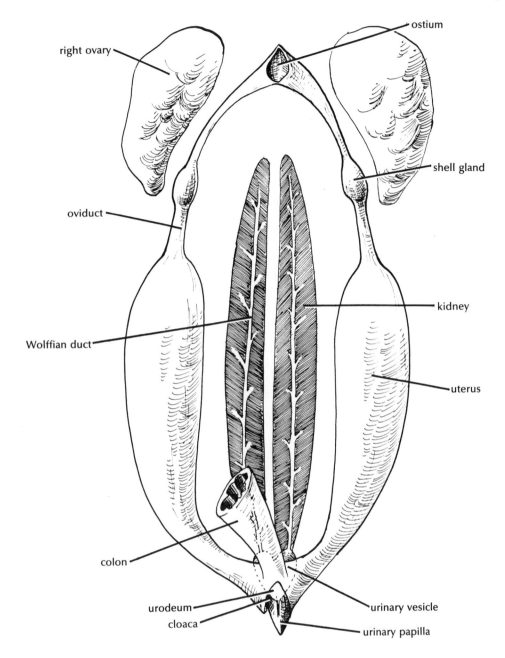

FIGURE 7.4. Schematic diagram of the female urogenital system, ventral view

duct may be observed by cutting through the parietal peritoneum along the lateral border of the kidney and reflecting it gently from the kidney surface. Trace it caudally and note that it unites with the Wolffian duct of the opposite side to form an enlarged chamber called the **urinary vesicle** before joining the cloaca.

OVARIES

The ovaries are a pair of large organs adjacent to the cranial ends of the kidneys, although in immature specimens they are relatively small and smooth in texture.

They are the female gonads; as such, they produce the female gametes, or **ova**, and female sex hormones. In some specimens the ova, packaged within follicles, may be visibly present as internal oval nodules pushing against the ovary wall.

OVIDUCTS

The oviducts are paired tubes located medial to each ovary and suspended in the pleuroperitoneal cavity by the **mesotubarium**. In immature specimens, they are small and lack the mesotubarium. Each originates at its cranial end within the falciform ligament, where it unites with the oviduct of the opposite side. At this union, the oviducts share a common opening, called the **ostium**, which opens into the pleuroperitoneal cavity through a small slit. In many specimens, this is difficult to find.

From the ostium, the oviduct extends caudally along the ventral surface of the kidney before it enlarges at a point dorsal to the ovary. This enlargement is the **shell (nidamental) gland**, which secretes a thin shell around groups of two or three eggs as they descend the oviduct. Continuing caudally from this gland, the oviduct becomes narrow once more until it unites with the **uterus**.

UTERUS

The paired uteri of the dogfish are enlargements of the caudal half of each oviduct. They extend caudally to ultimately unite with the cloaca. These organs are quite large in pregnant females, for it is here that *gestation*[1] occurs. Gestation in the dogfish is quite surprising for it lasts twenty to twenty-two months. During this entire period, the dogfish embryos are nourished by attached internal and external yolk sacs. Once gestation is complete, they

---

[1] *Gestation* is the internal incubation of developing young.

are released to the external world as self-sufficient individuals. This type of development is called *ovovivipary*.[2] If your female specimen is pregnant, make an incision through the uterine wall and examine the embryos within.

CLOACA

The female cloaca is a receptacle for sperm from the male during copulation, urine from the **urinary vesicle**, solid waste material from the rectum, and the passage of dogfish "pups" during birth. Open the cloaca by making an incision through the cloacal aperture along one side and extend it into the lateral wall of the rectum. Note the entrance of the two uteri into the caudal portion of the cloaca at a point ventral to the **urinary papilla**. Note also the presence of a urogenital portion, called the **urodeum**, which is separated from the fecal portion by a horizontal fold.

---

[2] There are some species of sharks that receivenourishment and oxygen from the mother via the placental capillary network, which is similar to placental mammals. This is termed *vivipary*. Most other fishes, as well as amphibians, reptiles, birds, and monotreme mammals, release eggs into the external environment. This is termed *ovipary*.

The Nervous System & Special Senses

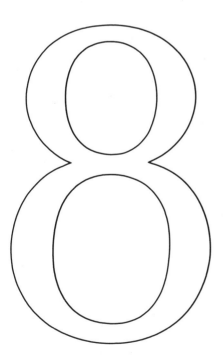

THE PRIMARY FUNCTION of the nervous system as a whole is to maintain the body in a relatively stable condition, or **homeostasis**, despite fluctuating environmental conditions. This is accomplished by controlling body activities through the generation of nerve impulses that travel along the nerve cells or **neurons**. This means of body control is rapid and specific to the region that receives the impulse.

In all vertebrates, the nervous system is divided into a **central nervous system (CNS)** and a **peripheral nervous system (PNS)**. The CNS consists of the **brain** and **spinal cord** and serves as the information input and output control center for the body. The PNS is composed of neurons arranged in bundles called **nerves**, which extend between the CNS and the peripheral regions of the body. PNS nerves transmit information from neurons that initiate an impulse when stimulated, called **receptors**, to the CNS and from the CNS to muscles and glands, called **effectors**. Some receptors are highly specialized in structure and function to respond to a particular stimulus. These receptors are called **special sensory organs**.

In this chapter, the components of the central nervous system and the special sensory organs that can be examined through dissection are presented. The peripheral nervous system is discussed in part only; the major peripheral nerves will not be considered because of the difficulty encountered in their dissection.

THE BRAIN

The brain of the dogfish represents an excellent example of a primitive vertebrate brain. Its major divisions, all of which may be compared to those of higher vertebrates, are within reach of dissection. These divisions are indicated in Figure 8.1 and include the **prosencephalon**, the **mesencephalon**, and the **rhombencephalon**. In addition, the prosencephalon is subdivided into the **telencephalon** and the **diencephalon**, and the rhombencephalon is subdivided into the **metencephalon** and the **myelencephalon**. Identify the visible components of these divisions, using Figure 8.1 before proceeding. In the dissection pro-

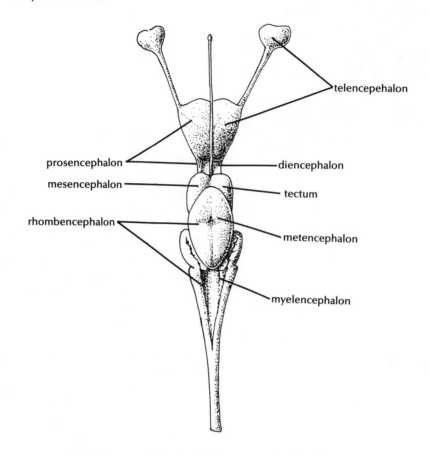

FIGURE 8.1.   The major divisions of the brain, dorsal view

tocol below, you will examine the structures and divisions of the brain in more detail, using a regional approach. For ideal study, the head of a large specimen that has not been used in previous dissections is suggested. It is also best to use separate heads for the study of the dorsal and ventral aspects if possible.

DORSAL ASPECT

Remove the skin and other tissue from the dorsal surface of the head and around the eye. Carefully cut away the dorsal portion of the chondrocranium to expose the cranial cavity. As you near the rostral part of the cranial cavity, note the presence of a slender stalk that extends from the center of the brain to a hole in the roof of the cranial cavity. This is the **epiphysis** or **pineal body**, which is an extended part of the diencephalon. Continue to expose the brain by removing the supraorbital crest and as much of the lateral walls of the cranial cavity as is reasonable without damaging the cranial nerves. Identify the following brain structures that are now visible from the dorsal view (Fig. 8.2):

Telencephalon: the anterior-most region of the brain, it consists of the **olfactory bulbs**, **olfactory tracts**, and the **cerebral hemispheres**.

Olfactory bulb (paired): enlargements at the anterior extremity located on the medial wall of each **olfactory sac**. The olfactory bulbs receive olfactory (sense of smell) input from cranial nerve I.

Olfactory tract (paired): a stalk on each side that serves as a bridge between the olfactory bulbs and the cerebral hemispheres.

Cerebral hemispheres (paired): oval structures that form the anterior segment of the main body of the brain. They are not highly developed as they are in higher vertebrates and receive only a limited input of sensory information through the thalamus (described below).

Diencephalon: a constricted region posterior to the cerebral hemispheres. It consists of the **epithalamus**, **thalamus**, and **hypothalamus**. The hypothalamus cannot be observed from the dorsal aspect.

Epithalamus: forms the roof of the diencephalon. A part of it is formed by a thin membrane called the **tela choroidea**. Carefully remove this membrane and observe the cavity that lies deep to it, called the **third ventricle**. A large fold that extends from the tela choroidea into the third ventricle represents the anterior end of the diencephalon. This fold is the **velum transversum**.

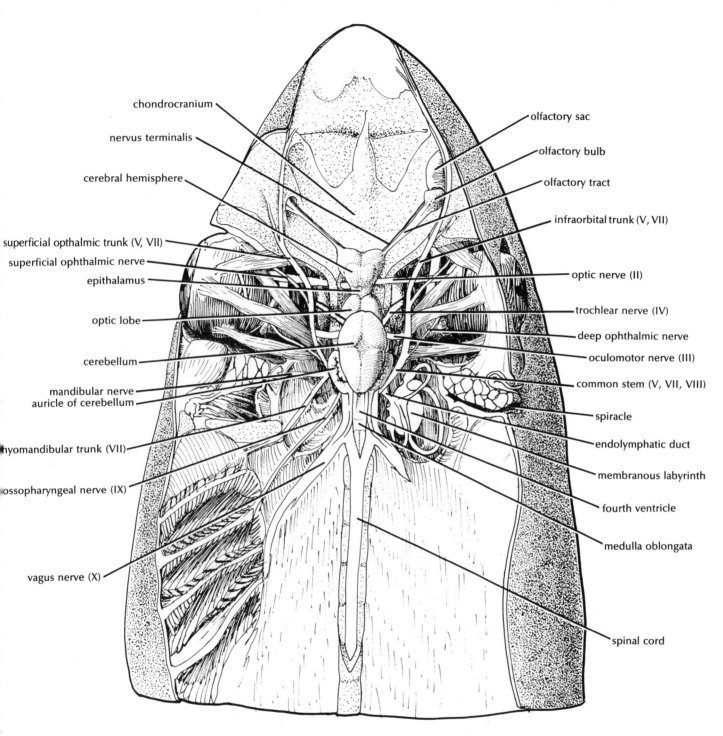

**FIGURE 8.2.   The chondrocranium with the brain and cranial nerves exposed, dorsal view.
The dissection has proceeded slightly deeper on the right side.**

Thalamus: forms the lateral walls of the third ventricle. The thalamus relays impulses to and from the cerebral hemispheres and other areas of the brain.
Mesencephalon: dorsally represented by the paired **optic lobes**.

Optic lobes (paired): large swellings posterior to the diencephalon that form the roof, or **tectum**, of the mesencephalon. The optic lobes receive visual input from the eyes. They also receive sensory information from other sensory centers of the brain and transmit motor impulses outward, thereby representing the master integration center that is analogous to the cerebral cortex of higher vertebrates.

Metencephalon: on the dorsal side, it is represented by the **cerebellum**.

    Cerebellum: a large, oval structure that partially overhangs the optic lobes at its anterior end. Notice the shallow grooves on its surface, which subdivide the cerebellum into four parts. On either side of its posterior half are located lateral projections called **auricles of the cerebellum**. The cerebellum is the center for muscular coordination and receives sensory fibers from the inner ear and lateral line system.

Myelencephalon: consists of the **medulla oblongata** and is the posterior segment of the brain.

    Medulla oblongata: located between the cerebellum and spinal cord, most of its roof is formed by a **tela choroidea** that overlies a cavity called the **fourth ventricle**. Remove the tela choroidea and examine the fourth ventricle. Observe two parallel ridges along its floor. These are the **somatic motor columns**, which are the ventral most of four longitudinal bundles of gray matter that pass through the medulla. Obtain a cross section of the medulla and identify the **visceral motor column** located lateral to the somatic motor columns, the **visceral sensory column** dorsal to the visceral motor columns, and the dorsally located **somatic sensory column** immediately deep to the tela choroidea attachments (Fig. 8.3). Each of these columns contain cell bodies of neurons that carry impulses in the direction indicated by their descriptive name. These columns are continuous posteriorly with the spinal cord and continue anteriorly through the meten-cephalon and into the mesencephalon, where they fragment into nuclei of gray matter.

CRANIAL NERVES

The organization of cranial nerves in the dogfish provides an excellent example of nervous design that is common among higher vertebrates. They provide direct pathways of communication between body structures and the brain and may be **motor (efferent)**, **sensory (afferent)**, or **mixed** (motor and sensory) in their nature of impulse conduction.

There are eleven pairs of cranial nerves in the dogfish. These are described below and summarized in Table 1. Note that most cranial nerves may be located near their attachment to the brain and, therefore, may be identified on the dorsal, ventral, and lateral aspects of your specimen's brain. As you study the following descriptions of the cranial nerves, refer to Figures 8.2, 8.4, and 8.5, and trace the nerves as far as possible in your specimen.

0. Nervus terminalis. a nerve that extends from the longitudinal fissure to the olfactory bulb. It is visible as a slender strand on the medial side of the olfactory tract. Its function is not well understood.

1. Olfactory nerve (I). a sensory nerve that consists of neurons originating from the olfactory epithelium of the olfactory sac. These neurons conduct impulses to the olfactory bulb, where they terminate. The olfactory tract conveys the impulse from the bulb to the cerebral hemisphere.

2. Optic nerve (II). a sensory nerve that conveys impulses from the retina of the eye to the optic chiasm at the base of the diencephalon.

3. Oculomotor nerve (III). a motor nerve that originates from the floor of the mesencephalon at a point dorsal to

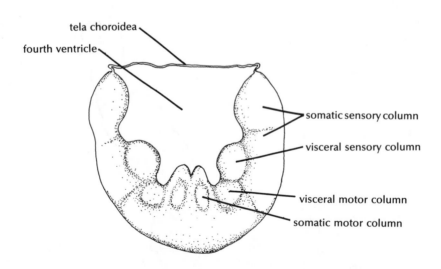

FIGURE 8.3. Transverse section through the medulla oblongata

TABLE 1. The Cranial Nerves

Cranial Nerve	Point of Union with Brain	Innervation	Sensation/Action
Nervus T. (0)	Longitudinal fissure	Unknown	Unknown
Olfactory (I) (sensory)	Olfactory lobe	Olfactory epithelium of nasal cavity	Sense of smell
Optic (II) (sensory)	Optic chiasm	Sensory cells in retina of eye	Sense of vision
Oculomotor (III) (motor)	Ventral aspect of mesencephalon	Intrinsic and most extrinsic eye muscles	Movement of eye, control of light entering eye
Trochlear (IV) (motor)	Dorsal aspect of mesencephalon between the optic lobes and cerebellum	Superior oblique muscle of the eye	Rotation of eyeball
Trigeminal (V) (mixed)	Ventral aspect of medulla, in common with VII and VIII	Skin of head region and muscles of first visceral arch	Sensation of skin at head region; jaw movement
Abducens (VI) (motor)	Ventral aspect of medulla posterior to common origin of V, VII, and VIII	Lateral rectus muscle of eye	Movement of eyeball
Facial (VII) (mixed)	Ventral aspect of medulla, in common with V and VIII	Ampullae of Lorenzini; lateral line of head, and mouth; muscles of second visceral (hyoid) arch	Senses of vibrational movement and taste; jaw movement
Auditory (VIII) (sensory)	Ventral aspect of medulla, in common with V and VII	Sensory regions of the membranous labyrinth	Senses of hearing and equilibrium
Glossopharyngeal (IX) (mixed)	Ventral aspect of medulla posterior to common origin of V, VII, and VIII	Caudal lateral line system, pharyngeal muscles	Sense of vibrational movement and first branchial pouch; movement of third branchial arch
Vagus (X) (mixed)	Ventral aspect of medulla posterior to origin of IX, by various roots	Mouth, branchial pouches 2 to 5, lateral line; muscles of branchial arches 2 to 5, heart and cranial part of G.I. tract	Sense of touch and movement of innervated regions; movement of branchial arches 2 to 5; contraction of heart wall and cranial G.I. tract

the vascular sac. From here, it extends into the orbit and divides into dorsal and ventral branches. The dorsal branch innervates the superior and medial rectus muscles, and the ventral branch extends posteriorly to supply the inferior rectus muscle, then anteriorly to supply the inferior oblique muscle. The oculomotor nerve, therefore, plays a major role in controlling eye movement. A small tributary of its ventral branch conducts autonomic impulses to the smooth muscles of the iris and ciliary body. Called the ciliary nerve, it regulates the amount of light entering the eye and accommodation.

4. Trochlear nerve (IV). a motor nerve that emerges from the roof of the mesencephalon between the optic lobe and the cerebellum. From here it extends anteriorly to enter the orbit near the origins of the superior rectus and superior oblique muscles. It supplies the superior rectus muscle and thereby helps control eye movement.

5. Trigeminal nerve (V). a mixed nerve that arises from a large common stem at the anterior end of the medulla. This stem carries with it cranial nerves V, VII, and VIII. Near its origin, the stem gives off VIII, which extends to the membranous labyrinth. From this division the common stem, which at this point contains V and VII, extends through the orbit wall and divides into six

branches. Two of these branches (called the **superficial ophthalmic trunk** and the **infraorbital trunk**) contain components of both cranial nerves V and VII; two branches are composed exclusively of fibers of VII, and two branches consist only of fibers of the trigeminal nerve (V) (called the **deep ophthalmic** and **mandibular nerves**). Thus, the trigeminal nerve has a total of four branches: the **superficial ophthalmic**, which is part of the superficial ophthalmic trunk; the **deep ophthalmic**; the **maxillary**, which is part of the infraorbital trunk; and the **mandibular**.

The superficial ophthalmic nerve originates as a branch from the superficial ophthalmic trunk within the orbit. It gives off numerous branches, which pass through foramina in the roof of the orbit and extend to the rostrum. Its trigeminal component conveys sensory impulses from the skin in the region dorsal to the orbit.

The deep ophthalmic nerve originates directly from the common stem and extends laterally to the medial aspect of the eyeball. From here, it makes a medial turn to pass ventral to the superior oblique muscle. It then continues anteriorly to the rostrum, which it supplies with sensory fibers.

The maxillary nerve extends as part of the infraorbital trunk through the anterior wall of the orbit. It divides

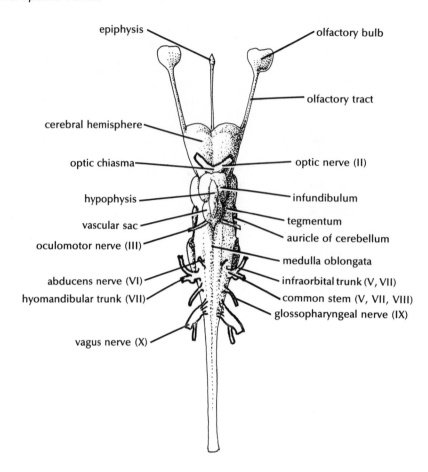

FIGURE 8.4.   The brain, ventral view

into branches that continue to the ventral side of the rostrum. The maxillary nerve supplies sensory fibers to the skin of the rostrum.

The mandibular nerve extends between the eye and the otic capsule before it turns ventrally to the muscles of the jaw. It supplies sensory fibers to the skin of the lower jaw and motor fibers to the muscles of the first visceral arch. It thereby controls opening and closing of the mouth.

6. **Abducens nerve (VI).** a motor nerve that arises from the ventral side of the medulla near the midline. It extends in an anterior direction along the common stem of nerves V, VII, and VIII to the orbit. It innervates the lateral rectus muscle.

7. **Facial nerve (VII).** a mixed nerve that consists of four branches: a **superficial ophthalmic**, which is part of the superficial ophthalmic trunk; a **buccal**, which is part of the infraorbital trunk; a **palatine**; and a **hyomandibular**.

The superficial ophthalmic nerve follows the same route as described under "trigeminal." Its facial nerve component supplies sensory fibers to the lateral line system and ampullae of Lorenzini in the region dorsal to the eye.

The buccal nerve is the facial nerve component to the infraorbital trunk, which it shares with the maxillary nerve of the trigeminal. The buccal nerve contains sensory fibers that supply the lateral line system and ampullae of Lorenzini in the region ventral to the eye.

The palatine nerve arises from the short trunk that is in common to the hyomandibular. It extends anteriorly to the mouth. The palatine supplies sensory fibers to the taste buds and epithelium of the mouth.

The hyomandibular nerve originates from a short trunk that it shares with the palatine nerve. From its origin, the hyomandibular extends posteriorly to the otic capsule and emerges between muscles surrounding the spiracle. Its sensory fibers supply the tongue, floor of the mouth, and the lateral line system and ampullae of Lorenzini in the region of the mouth. Its motor fibers innervate the muscles of the second visceral arch, and it therefore plays a role in jaw movement.

8. **Auditory nerve (VIII).** a sensory nerve that originates from the common stem of cranial nerves V, VII, and VIII. From here it extends to the otic capsule, where it divides into branches that continue to the membranous labyrinth. It conveys sensory information (for sound and equilibrium awareness) to the medulla.

9. **Glossopharyngeal nerve (IX).** a mixed nerve that arises from the medulla posterior to the common stem of cranial nerves V, VII, and VIII. It extends posteriorly

from its origin toward the first branchial pouch. The glossopharyngeal nerve supplies sensory fibers to the first branchial pouch and pharynx and motor fibers to the first branchial pouch and third branchial arch by way of three smaller branches (pretrematic, posttrematic, and pharyngeal branches).

10. **Vagus nerve (X)**. a large mixed nerve that arises from the lateral aspect of the medulla by way of numerous rootlets. From its origin posterior to the origin of the glossopharyngeal nerve, it extends toward the branchial pouches and enters the anterior cardinal sinus. Within the anterior cardinal sinus, the vagus nerve divides into a prominent **visceral branch** and a smaller **lateral branch**. From the visceral branch arise four **branchial nerves**, which extend to branchial pouches and branchial arches two through five. The branchial nerves contain both sensory and motor fibers, which supply these branchial regions. The visceral branch then continues posteriorly to provide the heart and the cranial part of the G.I. tract with autonomic sensory and motor fibers. The lateral branch of the vagus nerve supplies much of the lateral line system with sensory fibers.

VENTRAL ASPECT

To study the brain from its ventral side, you must remove it from the cranial cavity. Begin this procedure by cutting across the posterior end of the medulla and the anterior end of the olfactory tracts. Next cut through the cranial nerves, but leave as much nerve material attached to the brain as possible to allow them to be identified. Carefully push the brain to one side and cut through the small nerve that extends ventrally from the underside of the brain. This is the **abducens nerve**. To avoid tearing the hypophysis when removing the brain, loosen its connections to the chondrocranium. The brain should now be free of attachments except for connective tissue. Remove the brain from the cranial cavity by lifting the posterior end first and pulling gently forward. If you encounter resistance, tease it free with a dull probe.

Once the brain is removed, identify the five divisions previously observed from the dorsal aspect, and locate the following structures from the ventral view (Fig. 8.4):

Telencephalon: Locate once more the **olfactory tracts** and the **cerebral hemispheres** (the **olfactory bulbs** were removed during dissection).

Diencephalon: The **epithalamus** and **thalamus** are not visible from the ventral view, but the **hypothalamus** and its associated structures may be clearly observed.

    Hypothalamus: The structure forming the floor of the diencephalon, it is posterior to the cerebral hemispheres. It is the largest portion of the diencephalon and contains important visceral and autonomic centers. These centers, which are represented by nuclei of gray matter, regulate digestion, metabolic activity, blood sugar level, water balance, and sex-

ual activity. The paired **optic nerves** attach at its anterior end where they cross (or *decussate*), forming an X- shaped structure called the **optic chiasma**. In dogfish as in other vertebrates but mammals, decussation is complete in that fibers from one eye travel to the optic lobe on the opposite side of the brain. The remaining ventral surface of the hypothalamus is mainly occupied by the **infundibulum**, which consists of a pair of **inferior lobes** divided medially and a thin-walled sac posterior to the inferior lobes called the **vascular sac**. Between the inferior lobes along the midline and ventral to the vascular sac is an important endocrine gland termed the **hypophysis** or **pituitary gland**.

Mesencephalon: The floor of this region is visible dorsal to the hypophysis and is called the **tegmentum**.

Metencephalon: located posterior to the tegmentum. The lateral **auricles of the cerebellum** are also visible ventrally.

Myelencephalon: The ventral surface is visible posterior to the metencephalon. There is no line of division between these two regions.

SAGITTAL ASPECT

Place the brain of your dogfish specimen so that it can be viewed laterally, and identify the structures from this view that have been studied from the dorsal and ventral aspects (Fig. 8.5). Next, place the brain ventral side down on a wax tray. With a single-edged razor blade or large scalpel, cut the brain in right and left halves along the midline, beginning from the medulla and proceeding anteriorly. Try to make this cut with a single stroke to prevent damage to internal structures. Keep the brain halves moist as you identify the following structures (Fig. 8.6):

Telencephalon: Locate the cerebral hemispheres. Each hemisphere contains a cavity, the **lateral ventricle**, which is continuous with the central cavity of the olfactory bulb and tract. Within the lateral ventricles, **cerebrospinal fluid (CSF)** is formed by filtration through a rich supply of capillaries called the **choroid plexus**. The lateral ventricles represent the first two of six cavities in the brain through which CSF circulates. Locate the **foramen of Monro**, which allows passage of CSF from the lateral ventricles to the **third ventricle** in the diencephalon.

Diencephalon: Note the location of the **third ventricle**, which separates the two portions of the thalamus.

    Epithalamus: Observe the location of the **tela choroidea** and the **velum transversum** in the sagittal view. Posterior to the tela choroidea is a small mass of nervous tissue, the **habenula**, which attaches to the long, narrow **epiphysis** that extends anteriorly. The habenula is an olfactory center in the dogfish. Posterior to the habenula is the **posterior commissure**, which is a tract of white matter bridging

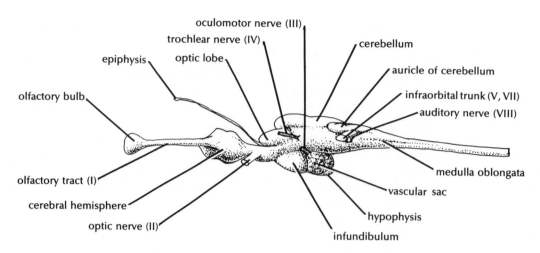

FIGURE 8.5.   The brain, lateral view

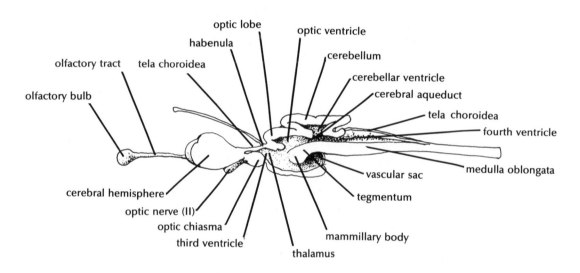

FIGURE 8.6.   Midsagittal section of the brain, lateral view

the opposite sides of the brain.

Thalamus: visible as an area of gray matter in the lateral wall of the third ventricle.

Hypothalamus: Compare the **optic chiasma**, the **hypophysis**, and the **infundibulum** in the sagittal view with the dorsal and ventral views studied earlier. Locate also the **mammillary body**, which forms the thickened floor of the diencephalon just above the vascular sac and was not visible in other views.

Mesencephalon: locate the **optic lobe**, and note that it contains a cavity. This space is the **optic ventricle**, which communicates with the **cerebral aqueduct**. The cerebral aqueduct channels CSF from the third ventricle to the fourth ventricle located within the myelencephalon.

Metencephalon: Locate the **cerebellum** in the sagittal aspect. The cavity within is called the **cerebellar ventricle**, which communicates with the cerebral aqueduct.

Myelencephalon: Locate the medulla oblongata once more, and identify the cavity deep to the tela choroidea as the **fourth ventricle**. On the lateral wall of the fourth ventricle, observe the **somatic sensory column** and the **visceral sensory column** described earlier. From the fourth ventricle, CSF exits the brain through small openings (**apertures**) to continue its circulation around the exterior surface of the brain and spinal cord and through a small canal in the center of the spinal cord before being reabsorbed into the bloodstream.

THE SPINAL CORD

The spinal cord lies within the vertebral canal of the vertebral column. At its anterior end it is continuous with the medulla, and it extends posteriorly to terminate near the tip of the tail. To view the spinal cord and its associated structures, place your specimen on its ventral side and remove the muscles overlying the vertebral column of the trunk over a region about five cm in length. Carefully cut away the dorsal and lateral walls of the vertebral arch while avoiding the white strands of nerve tissue. From above, observe the white **spinal cord** and the **dorsal roots** of spinal nerves. Now obtain a cross sectional segment of the cord by cutting across it at two points about two cm apart through a region where the vertebral arch has not been removed. Thus, you will be cutting through the vertebra and the cord completely. Cut all attachments within one cm from the vertebral segment with the cord remaining intact inside, and remove it. Identify the components of the spinal cord, as shown in Figure 8.7.

Gray matter: In a stained cross section, it may be observed to occupy the central region of the cord. Here, it is divided into a **dorsal horn** and a **ventral horn**. The gray matter consists of cell bodies of neurons and their dendrites.

White matter: consists of myelinated axons of fibers passing up and down the cord, called **tracts**. The fiber tracts are arranged in bundles, or **funiculi**, which occupy the peripheral regions of the spinal cord.

Central canal: a small canal in the center of the spinal cord that contains CSF and is continuous with the fourth ventricle.

Dorsal root: the dorsal attachment of each spinal nerve to the spinal cord, which exits from the vertebral canal through foramina in the vertebral arch. The dorsal root consists of afferent (sensory) fibers and visceral motor fibers. It contains a swelling near its union with the spinal nerve, called the **dorsal root ganglion**, which consists of afferent neuron cell bodies.

Ventral root: the ventral attachment of each spinal nerve to the spinal cord. It consists of only efferent (motor) fibers. Like the dorsal roots, the ventral roots also exit from the vertebral canal through foramina to unite to form each spinal nerve.

Spinal nerve: mixed nerves containing afferent and efferent fibers that arise from the spinal cord via dorsal and ventral roots. The two roots emerge from the cord in alternating fashion, such that the ventral roots are aligned with myomeres, and the dorsal roots extend between them. They represent the origin or termination of peripheral nerves that provide communication between the skin, muscles, and viscera and the central nervous system.

The **autonomic nervous system**, which controls involuntary activities in vertebrates, will not be studied in the dogfish because of the difficulty of dissection. It is represented by the presence of preganglionic fibers that pass between the spinal cord and ganglia located in the posterior cardinal sinus and the kidney, and postganglionic fibers between the ganglia and the tissues of the kidney and G.I. tract.

ORGANS OF SPECIAL SENSE

The organs of special sense are sensory structures that are specialized to respond to a particular type of stimulus. The result of this response is the generation of a nerve impulse that, when received by the brain, may be interpreted as a sensation. The organs of special sense that will be considered are the **lateral line system**, the **ear**, the **olfactory organ**, and the **eye**.

LATERAL LINE SYSTEM

The lateral line system is a series of small canals that form a line on the lateral surface of the trunk and a series

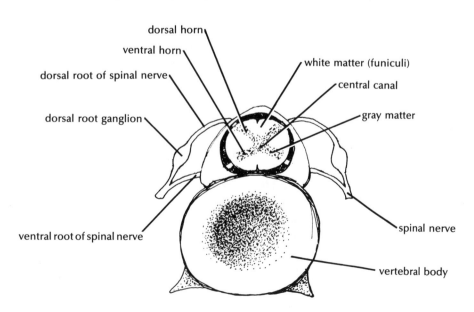

FIGURE 8.7.   The spinal cord and related structures, transverse section

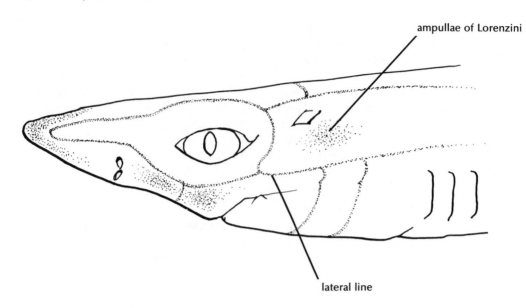

ampullae of Lorenzini

lateral line

FIGURE 8.8.   The lateral line system of the head region

of lines on the head of fishes and larval amphibians (Fig. 8.8). Within the canals are sensory cells that respond to mechanical disturbances in the water, thereby providing the fish with the ability to sense its own movement and that of the environment around it. The sensory cells that respond to this movement are called **neuromasts**.

Observe the lateral line system on the dogfish by examining the skin on top of the head, around the eyes, and beneath the snout. The canals that form the lateral line are located immediately beneath the skin, and their distribution is portrayed by the presence of small openings along short intervals that permit water passage. The lateral line canals are named according to their locations relative to the orbit and mandible.

In addition to the lateral line canals are numerous pores that release a gelatinous exudate when squeezed. These pores are the entrances to organs associated with the lateral line system called **ampullae of Lorenzini**. They are numerous on the ventral surface of the head, around the snout, and behind the eyes. They respond to changes in water pressure, salinity, and temperature, possibly through electroreception.

THE EAR

The ear is the organ of hearing, or **audation**, and balance, or **equilibrium**. It is closely related to the lateral line system, as the receptor cells of the ear are very similar in structure and function to neuromasts. The ear of fishes consists of only an inner portion, which functions in equilibrium (body position) as well as in hearing. In terrestrial vertebrates, the ear consists of outer, middle, and inner portions.

The inner ear of the dogfish consists of a fluid-filled membranous structure composed of tubes and chambers

called the **membranous labyrinth**, locked within a cavity inside the otic capsule and bordered by cartilaginous walls. To dissect it into view in your specimen, first locate the two **endolymphatic ducts** near the dorsal midline between the spiracles. These ducts are openings into the membranous labyrinth that permit water movement between it and the exterior. Scrape the skin away from this region and observe an endolymphatic duct passing deep toward the inner ear. Slice the cartilage away in thin sections until you can see the inner ear through its transparency. Begin from the dorsal and lateral surfaces of the otic capsule, and proceed ventrally. Without dissecting further, compare the structure of the inner ear with Figure 8.9, and identify the following:

Cartilaginous labyrinth: the cavity within the otic capsule that contains the membranous labyrinth. The space between it and the membranous labyrinth is filled with a fluid called **perilymph**.

Membranous labyrinth: the membranous structure composed of tubes and chambers. It is filled with a fluid called **endolymph**. It consists of the following components:

    Semicircular ducts: three curved, narrow channels that lie anterior, posterior, and horizontal. At the ventral end of each duct is an expanded area called the **ampulla**, which contains a sensory region known as the **crista**.

    Sacculus: the large central chamber that receives the endolymphatic duct. The small protrusion from its caudoventral region is the **lagena**. Within the sacculus is a large mass of calcareous grains called **otoliths**.

    Utriculus: smaller chambers that continue from the semicircular ducts and fuse with the sacculus.

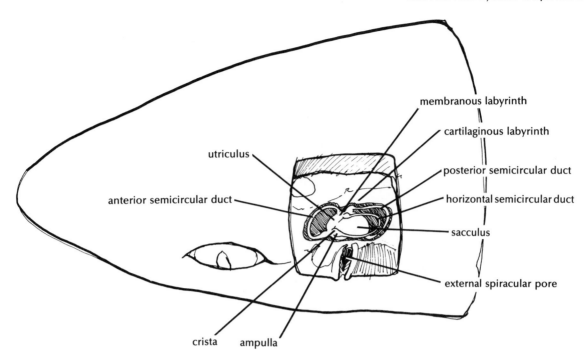

FIGURE 8.9.   The inner ear apparatus

OLFACTION

The sense of smell, or **olfaction**, is localized within the **olfactory sacs** in the dogfish. Identify the following olfactory structures in your specimen (Fig. 8.10):

Nares: the openings in the snout which permit water flow between the olfactory sacs and the exterior. Each naris is divided into two channels by a septum; the lateral channel is the incurrent opening and the medial channel is the excurrent opening.

Olfactory sac: Remove the skin and cartilage from around the side of the snout to expose the olfactory sac on one side. Section the round sac, and notice its internal divisions into numerous folds of tissue called **olfactory lamellae**. The lamellae contain receptor cells sensitive to chemical stimuli. Their fibers pass to the **olfactory bulb**, where they synapse with fibers that extend through the **olfactory tract** to terminate in the cerebral hemisphere.

THE EYE

The eye is the organ of **vision** and lies partially embedded within the orbit of the chondrocranium. Examine the eye of your specimen and notice the **upper** and **lower eyelids**. These immovable skin folds are not present in most other fishes. Near the junction between the lid and the eyeball is the **conjunctiva**, which is a continuation of epithelium from the lid that partially covers the eyeball. Now cut away the cartilage and other tissues that surround the orbit, and carefully pull the eyeball partially

out of the orbit. While the eyeball remains attached, identify the following six **extrinsic muscles** and other structures that lie between it and the posterior wall of the orbit (Fig. 8.11).

Superior rectus muscle: extends between the posterior wall of the orbit and the superior surface of the eyeball. It is innervated by the oculomotor nerve (III).

Inferior rectus muscle: extends between the orbit wall and the inferior surface of the eyeball. It is innervated by the oculomotor nerve (III).

Lateral rectus muscle: extends between the orbit wall and the lateral surface of the eyeball. It is innervated by the abducens nerve (VI).

Medial rectus muscle: extends between the orbit wall and the medial surface of the eyeball. It is innervated by the oculomotor nerve (III).

Superior oblique muscle: originates from the anterior medial corner of the orbit and passes to the superior eyeball surface. It is innervated by the trochlear nerve (IV).

Inferior oblique muscle: extends between the anterior medial corner of the orbit to the inferior surface of the eyeball. It is innervated by the oculomotor nerve (III).

Optic nerve: a thick stalk that attaches the posterior wall of the eyeball to the orbit. It is composed of nerve fibers that originate within the retina of the eye and pass to the thalamus.

Optic pedicle: a cartilaginous extension from the posterior orbit wall between the four rectus muscle origins. It projects in an anterior direction to the back of the eye-

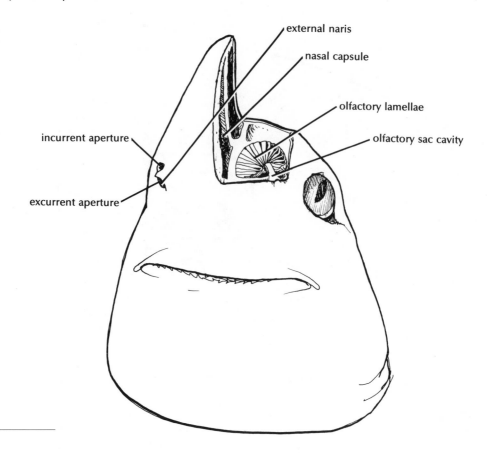

FIGURE 8.10.   The structures of olfaction

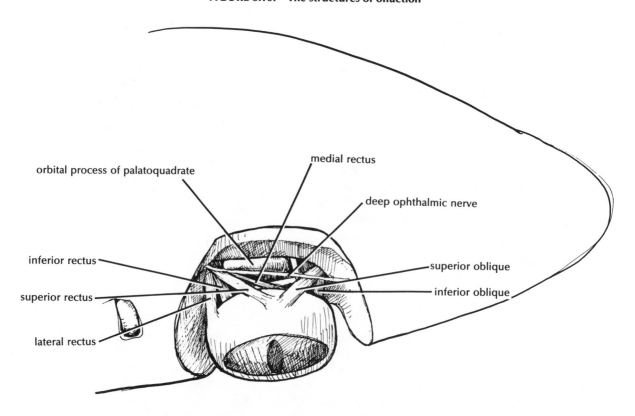

FIGURE 8.11.   The extrinsic muscles of the eyeball, dorsal view

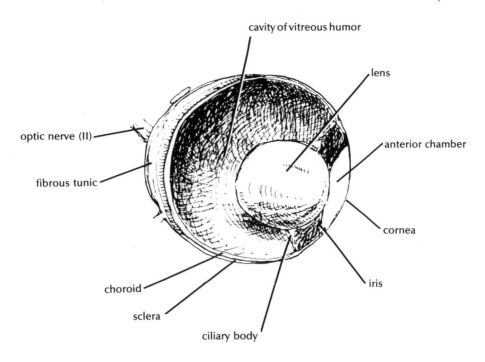

FIGURE 8.12. The eyeball, sagittal section

ball. Its anterior end contains a disk that provides support for the eyeball. This structure is not present in most vertebrate eyes.

Cut all attachments to the eyeball and remove it from the orbit intact. Place it on a dissecting tray and, with a sharp blade, cut it in half along its midsagittal plane. The eyeball is composed of three layers or **tunics**: an outer **fibrous tunic**, a middle **vascular tunic**, and an inner **sensory tunic**. Identify these layers and their components on the dissected eyeball (Fig. 8.12):

Fibrous tunic: the outer layer of the eye. It consists of the **sclera** and the **cornea**.

   Sclera: the posterior portion of the fibrous tunic. It is composed of white fibrous connective tissue and thereby provides a tough outer covering for the eyeball.

   Cornea: the transparent, anterior portion of the fibrous tunic. It is continuous with the sclera along its posterior edges.

Vascular tunic: the middle layer of the eyeball. It is a pigmented region that consists of the **choroid**, the **ciliary body**, and the **iris**.

   Choroid: the posterior segment of the vascular tunic. It is a heavily pigmented layer of vascular connective tissue that provides nourishment to the avascular **retina**, which lies immediately deep to it.

   Ciliary body: Located anterior to the choroid, it is a thin, black structure characterized by radial lines. Connective tissue and smooth muscle fibers extend from it to provide support for the **lens**. The lens in

the dogfish is nearly spherical and is normally transparent in life.

   Iris: Located between the cornea and the lens, it is the anterior segment of the vascular tunic. It is composed of smooth muscle fibers that regulate the amount of light entering the eyeball. The hole in the center through which light enters is the **pupil**.

Sensory tunic: Located deep to the choroid against the inner posterior wall of the orbit, it consists of the **retina**.

   Retina: A delicate white membrane that lines the inner posterior wall of the orbit, it can be easily separated from the choroid. It is composed of several layers of neurons, the most superficial of which consists of photoreceptive neurons. The fibers of the deepest layer exit the eyeball to form the **optic nerve**, which crosses at the **optic chiasma** before terminating at the **thalamus**. The point at which the optic nerve intersects the retina is the **optic disc**, which contains no photoreceptors and is therefore the "blind spot."

Identify the cavities of the eyeball in the sagittal section (Fig. 8.12). The large cavity posterior to the lens is called the **cavity of the vitreous humor**. It contains a gelatinous material called the **vitreous humor**, which aids in maintaining the internal pressure. The small cavity between the cornea and the iris is the **anterior chamber**, and the space between the iris and the lens is the **posterior chamber**. The anterior and posterior chambers contain a watery fluid that is constantly regenerated called the **aqueous humor**.

also by Bruce D. Wingerd and available from Johns Hopkins:

Human Anatomy and Rabbit Dissection
Rabbit Dissection Manual
Frog Dissection Manual
Rat Dissection Manual

The Johns Hopkins University Press

DOGFISH DISSECTION MANUAL

This book was composed in Optima (Oracle) type by
Brushwood Graphics, Inc., from a design by Susan P. Fillion.
It was printed by Thomson-Shore, Inc., on 60-lb. Spring
Forge Offset.